STORYTELLING APES

ANIMALIBUS VOL. 5
OF ANIMALS AND CULTURES

Nigel Rothfels and Garry Marvin, *General Editors*

ADVISORY BOARD:
Steve Baker *(University of Central Lancashire)*
Susan McHugh *(University of New England)*
Jules Pretty *(University of Essex)*
Alan Rauch *(University of North Carolina at Charlotte)*

Books in the Animalibus series share a fascination with the status and
the role of animals in human life. Crossing the humanities and the
social sciences to include work in history, anthropology, social and
cultural geography, environmental studies, and literary and art criticism,
these books ask what thinking about nonhuman animals can teach us
about human cultures, about what it means to be human, and about
how that meaning might shift across times and places.

OTHER TITLES IN THE SERIES:

Rachel Poliquin,
The Breathless Zoo:
Taxidermy and the Cultures of Longing

Joan B. Landes, Paula Young Lee, and Paul Youngquist, eds.,
Gorgeous Beasts:
Animal Bodies in Historical Perspective

Liv Emma Thorsen, Karen A. Rader, and Adam Dodd, eds.,
Animals on Display:
The Creaturely in Museums, Zoos, and Natural History

Ann-Janine Morey,
Picturing Dogs, Seeing Ourselves:
Vintage American Photographs

Storytelling Apes

PRIMATOLOGY NARRATIVES PAST AND FUTURE

Mary Sanders Pollock

The Pennsylvania State University Press
University Park, Pennsylvania

Library of Congress Cataloging-in-Publication Data

Pollock, Mary Sanders, 1948– , author.
Storytelling apes : primatology narratives past and future /
Mary Sanders Pollock.
pages   cm — (Animalibus)
Summary: "A literary analysis of the popular genre of the
informal primatology field narrative. Explores the works of
Jane Goodall, Dian Fossey, Robert Sapolsky, and others in
the contexts of scientific, literary, and conservation
discourses"—Provided by publisher.
Includes bibliographical references and index.
ISBN 978-0-271-06630-1 (cloth : alk. paper)
1. Primatology—Authorship.
2. Primatology—Fieldwork.
3. Creative nonfiction—History and criticism.
I. Title. II. Series: Animalibus.

QL737.P9P643   2015
599.8—dc23
2014043060

The Pennsylvania State University Press is a member of the
Association of American University Presses.

It is the policy of The Pennsylvania State University Press to use acid-free paper.
Publications on uncoated stock satisfy the minimum requirements of
American National Standard for Information Sciences—Permanence
of Paper for Printed Library Material, ANSI Z39.48–1992.

This book is printed on paper that contains
30% post-consumer waste.

FOR MY BROTHERS,

John, Tom, and Jim,

and

IN MEMORY OF

Ann Elizabeth Burlin

CONTENTS

ACKNOWLEDGMENTS

Like most other primates, we humans form social groups for safety, comfort, and fun. This book is the result of such a grouping—a "fission-fusion" group of friends and fellows. Without them, you would not be reading these stories.

After a campus visit by Jane Goodall, the late president of Stetson University, H. Douglas Lee, urged me to focus my animal studies research on primates. Soon after, I participated in a workshop at the National Humanities Center—and came away confident that I could do that. I am grateful to everyone at the NHC. I could not have continued without the help, conversations, and tolerance of my friends in the American Society of Primatologists, especially Evan Zucker and Karen Bales. At Stetson University, Dean Grady Ballenger and the many members of the Professional Development Committee over the years have generously supported this project with grant funding and sabbatical leave.

For reading and responding when I asked, I am deeply grateful to my department chairs Tom Farrell and John Pearson and to my colleagues Terry Farrell and Emily Mieras. Librarians Cathy Ervin and Susan Derryberry provided material support, Terry Grieb provided technical support, and Cathy Burke's help was essential in preparing this book for publication. Michelle Bezanson has contributed the wonderful line drawings at the beginning of each chapter; as well as anyone possibly could, she understands the borderland where art and science meet.

Without the support of Catherine Rainwater, Carrie Rohman, Karla Armbruster, and Sarah McFarland, I might have given up. The late Ann Burlin helped and encouraged me along the way. Thanks also to Patti Ragan at the Center for Great Apes. Rick Mueller, Nancy Vosburg, Ellen Phillips, Clifford Endres, Yves Clemmen, John Sanders, Tom Sanders, Marianne Sanders, and Vivian Sanders helped in many ways. Maggie the dog was supportive in her own way.

Without help from some of the scientists about whom I have written, I could not have completed this project. I am deeply grateful to Karen Strier for her advice and encouragement. Barbara Smuts and Robert Sapolsky responded generously to my requests for information and photographs. Lorna Joachim and Monica Mogilewsky opened their field sites and their hearts. I am grateful to Kendra Boileau and Julie Schoelles at Penn State University Press for understanding and honoring my intentions. I thank you all.

Introduction

We can no longer speak of reality . . . without considering how the world is altered and created when it is put into words.

—WALLACE MARTIN, *Recent Theories of Narrative*

I

I begin this story with a glance at some small monkeys in a small place and a scientist who loves them. The monkeys and the scientist are exemplary.

I look up from the treacherous clay path just as a toucan hops, high in the canopy, from one branch to another, its black silhouette and magnificent yellow beak clearly visible even from my position. "Are you sure it wasn't a monkey?" asks Lorna Joachim. "When they're going short distances, toucans sometimes hop like monkeys instead of flying." Almost as if invoked by our longing, monkey shapes now emerge from the leafy shadows, a hundred feet above our heads. They could be spider monkeys, capuchins, or howlers—three species who share this small forest fragment in northeast Costa Rica. Almost at once, Lorna identifies the shapes as mantled black howler monkeys, who are apparently deciding whether to travel on after napping or lunching.

Lorna sees monkeys in trees because she has a search pattern for the monkeys, consisting of type and speed of movement, size, coloration, location, and the company they keep. For this species, noise is the best identifying feature, but howlers howl only on their own schedules, or if they are protesting someone's presence. Since following howlers means constantly looking up through backlit branches, the slow, deliberate hopping movements of the monkeys are the first thing that can be spotted from the ground. Howlers are much heavier than the capuchins who also inhabit this area, and less agile than the resident spider monkeys, so they move more slowly. If the light is just right, an observer can see the males' fluffy white scrotal fur, and if the silhouette can be distinguished clearly from the background, head and body shape are also identifying features. Finally, these howlers typically live and travel in groups of fewer than ten. Their lives are so completely arboreal that, without the search pattern, the follower probably won't see the monkeys at all in their natural habitat.

While we watch, the small band of males, females (one with a clinging infant), and juveniles start to move gingerly from one branch to another and from one treetop to the next. Since I have never seen howlers except at the zoo, I am surprised by the care with which they pick their way. I had expected a rush of carefree abandon. But even though howlers have evolved to spend the majority of their time at the top of the forest, where the most tender leaves are found, monkeys do sometimes fall, and a fall from such a height would result in death or serious injury.

A single figure stands facing me, reaches up, and grasps the branch overhead, extending to full height. At such a distance, with the light filtering from above and behind, only the shape is visible: I cannot discern the color, sex, or size, although I can observe the monkey's careful movements, which suggest that it is relatively heavy and therefore an adult. In this dim and tricky light, my field glasses merely enlarge the silhouette. All I can see is that the body in the tree is like mine, even though it is doing something my body could have accomplished only during the most vigorous and adventurous years of my childhood. The monkey takes me up in space and back in time. It awakens my curiosity and my desire for understanding. I could fall in love with these beings and this life.

Lorna has. I witnessed her passion that very night as we shared our beer and played endless card games in the research center dining hall. Suddenly, semiautomatic weapon fire rang out a few yards away, and the howlers abruptly ended their evening canticle. Lorna's only fear was for the monkeys. The next day, after spending the night acting out the plot of a B movie—in a tiny SUV, on tooth-shattering roads, escaping to a sleazy motel twenty miles away—we found out that a local thug had been squatting in a cabin in La Suerte Bioreserve, which the owners have dedicated to conservation in return for a substantial tax reduction. The interloper was trying to lay claim to the cabin by means of an unusual (and irrelevant) Costa Rican law that discourages absentee ownership, and trying to impress the women he was entertaining with the size of his guns. This time, the monkeys were safe, and so were we. But all that is another story.

This book is about primates worldwide and the scientists who study them. Most primate species—apes, monkeys, and prosimians (lemurs and their close kin)—inhabit equatorial forests, uplands, and savannahs. For scientists from the so-called developed nations, these animals have been exotic, rare, and hard to study in the wild until the last fifty years, when postcolonial expansion opened up remote areas for communication and economic development. Tremendous economic, military, and environmental pressures followed, and many primate species are now on the brink of extinction before ever having been studied in their natural homes. In fact, as Jane Goodall explains in her most recent book, *Hope for Animals and Their World* (2009),[1] there is reason to believe that many species, including some primate species, could become extinct before they are even seen in the wild by primatologists.[2]

Fortunately, the small groups of howlers, capuchins, and spider monkeys observed by students in Lorna Joachim's field school belong to some of the best-known nonhuman primate species. Capuchins are the traditional "organ grinders' monkeys." Spider monkeys have long been zoo favorites. And howlers in Panama were first studied extensively in the 1930s by C. R. Carpenter, a student of Robert Yerkes, the founder of the discipline of primatology. Even though howlers, capuchins, and spider monkeys are not endangered, every small population is still important because, as habitats shrink and fragment, both the raw numbers and the genetic diversity of whole species decrease.

Throughout Cenozoic time, primates have developed into key players in tropical ecosystems as predators of insects and other small animals, seed dispersers, managers of undergrowth, and thinners of forest canopy. In addition, they make up a significant percentage of the forest biomass. Without primates, the forests that support them would not be themselves. And without the forests, nonhuman primates would be extinct except in zoos and research centers; they would not be themselves, either. Furthermore, almost every serious study of primates in the wild adds to the knowledge of the animals and, indirectly, of humans. Lorna and her students have been especially interested in observing play behaviors and trying to explain the capuchins' protocultural custom of lime washing: the monkeys scar the skin of a lime, briskly rub their fur with the fruit before discarding it, and start right away with a fresh lime. Although the result is a nice-smelling monkey (at least to human sensibilities) with some insect repellency, we can only surmise that a capuchin might engage in lime washing to get these results. The whole procedure might just be an adaptive accident.

We don't even know all the questions, much less the answers. Meanwhile, the monkey troops in La Suerte Bioreserve, Lorna's field site, could be lost to stray bullets shot into the air by show-offs or to the whim of the landlord who owns the reserve, which is not even a forest, but a collection of forest fragments.[3] There could be a change in the tax laws that now shelter the property as a reserve, a sale of the ranch to a banana exporter, a forest fire, an especially destructive hurricane, or some other consequence of global warming. If that loss were multiplied by a dozen, entire wild populations of these monkeys would indeed be at risk. Every field primatologist works on the cusp of a diminishing primate population and incalculable contingencies. Every one of them fears for the animals, with good reason. That is why I have written this book about the stories told by primatologists.

II

Humans are primates. We belong to the same order as singing siamangs, hamadryas baboons, and cotton-top tamarins. Genetically, we are more akin to the mouse lemur and slow loris than to the poodle sprawled on the carpet or the cat lounging on the kitchen table. We are apes. We share over 98 percent of our DNA with chimpanzees and bonobos—and it shows in many ways, not least in the uncanny oscillation between identification and repulsion that many people feel in their presence. On the basis of our shared genes and our shared evolutionary history, Jared Diamond calls humans "the third chimpanzee."[4] Apes, monkeys, and prosimians are significant in science and culture, not only because of the way they fill ecological niches but also because, more than any other animals, they serve as mirrors and surrogates for human beings.

My study focuses on primatologists from the West and the global North, whose work has been distinct from that of Asian primatologists. European and North American primatology developed in the early twentieth century from two different enterprises. The first was the parallel development of the social sciences *as* science, along with advances in medical science. Nonhuman primates have been considered for over a century to be the best models for studying the human body, human psychology, and human social behavior. In order to use nonhuman animals as models (with ample justification provided by philosophy and religion), Western science has typically defined nonhuman primates as similar to but not having the same transcendent significance as human beings. The second enterprise in the foundation of primatology was colonialism, which sent out European explorers, missionaries, and settlers to extract wealth in raw materials and knowledge from the exotic corners of the world—a process that has, in fact, only escalated since the official end of the colonial era. Of course, exploitation is not the purpose of scientific primatology. However, ironically, this background of European and neo-European expansion influences the shape of field narratives, which emphasize individual risk and discovery by primatologists, as well as ups and downs in the lives of the animals they study.

Japanese primatology has almost as long a history as Western primatology, but that tradition emerged from the study of Japan's own indigenous snow monkeys (Japanese macaques) and a focused interest on monkey behavior and society as a model for human culture. At the beginning of the twenty-first century, many other Asian countries, as well as countries in Latin America and Africa, now likewise have university primatology programs to train scientists to study their own indigenous primate species, but this development is relatively new. In these countries, the discipline has developed against deeply embedded archetypes of primates as gods, heroes, tricksters, and even human beings who perhaps took a different road sometime in the past. Although not the focus of this study, these ancient and new perspectives provide potentially rich threads if or when they are woven into Western understandings of primates.

The monkey god Hanuman, whose exploits are described in the Hindu epic *Ramayana,* is only one example of a primate archetype with a very long history, but I will mention him here because he figures in chapter 4. A henchman of the great King Rama, Hanuman is made a god for helping rescue Rama's wife, Sita, from the dark lord Ravana. Hanuman is kind, clever, brave, loyal, funny, and loquacious—and he can fly. He is very much the stereotype of the good monkey, except for the talking. (In one way of looking at it, some monkeys really can fly because they can speed through the treetops.) In a culture where primates have ontological status equivalent to humans, or in a country with indigenous primates, studying them is not considered exotic, and discoveries about primates are not necessarily feats of extreme individualism. Still, individualism and adventure are defining features of the field narratives written by Western primatologists—and that is one reason they are interesting to a broad readership in the West.

III

The literature of primatology includes academic peer-reviewed articles about experimental studies; books about field studies; and narrative accounts written for a mixed audience of scientists and lay readers. In (academic) scientific publications, technical language and the formulaic organization of material are designed to safeguard scientific accuracy and—perhaps equally important—the appearance of accuracy. Many field scientists find that the form cramps their style because it is wholly predictable, and the style sterilizes language against anthropomorphism—that is, the attribution of human qualities to nonhuman animals. Remove the scientist from the professional environment (in person or in print), and she typically refers to her study animals in thoroughly human terms, acknowledging with humor and irony that speaking of them in any other way is virtually impossible.

In contrast to publications written for fellow scientists, the storyworld that comes into being when a primatologist writes a field narrative—a literary zone somewhere between scientific argument and prose fiction—allows the lay reader to enter into the ordinarily formidable landscape of scientific discourse, while the scientist is allowed to speak in an authentic, personal voice. The field narrative is the focus of this book. The great contemporary novelist Ian McEwan gets the picture. "If one reads accounts of the systematic nonintrusive observations of troops of bonobo," he writes in *The Literary Animal*, a recent anthology of Darwinian literary criticism, "one sees rehearsed all the major themes of the English nineteenth-century novel: alliances made and broken, individuals rising while others fall, plots hatched, revenge, gratitude, injured pride, successful and unsuccessful courtship, bereavement and mourning."[5] In the same volume, E. O. Wilson speculates that the desire to replicate these plots is the result of the human mind's character as "a narrative machine, guided unconsciously by the epigenetic rules in creating scenarios and creating options."[6]

Collectively, the narratives under consideration here tell an additional story. Most of the scientists represented in this book are well-known figures in popular culture. Furthermore, although my selection is a fragment of the available field literature, these books, considered chronologically, reveal a history. They illustrate how the discipline of primatology—and the field as a site of knowledge production—has changed since the middle of the twentieth century. As a science, primatology has become more nuanced and necessarily more imbricated with the science of ecology. As a location, the field shrinks and decays with economic development, the expansion of human populations, war, and localized consequences of global warming. As the field and the discipline change, the narrative forms also change. If setting is an essential feature of most belletristic literature, so attention to geographical location is an essential feature of the primatology field narrative. At first, primatology narratives were about free-living animals whose lives had been virtually untouched by human activity; in 2015, most, if not all, primate populations are under threat, and reserves or sanctuaries are taking the place of the forest as field sites. As I show in this study, the shapes of the stories themselves evolve in response to changes in the setting/field.

My story begins before the earliest publication of primatology field narratives, however, with the Darwinian themes of evolution through natural and sexual selection and human kinship with other animals—themes that inform every one of the texts under consideration here. Darwin himself was a formidable storyteller, and like modern primatologists, he was fascinated by the behavior of apes and monkeys. After the 1859 publication of *The Origin of Species*, he began to write about them frequently. There are additional reasons why Darwin might be considered the first primatologist: he pondered the entire primate order and grappled with the evolutionary relationships among all the primate species he knew about. Apes and monkeys are so much like humans that dwelling on them in his first big book on evolution might have closed off escape routes for any reader who wanted to accept the theory of natural selection as it applied to other species but not to human beings. Darwin was sensitive to the dotted line that distinguishes the human from the animal and aware that most lay readers and many scientists saw this line as an absolute, divinely ordered boundary. Of course, by 1871, when Darwin published *The Descent of Man*, the cat was already out of the bag, so this book and later works are replete with references to primates.[7]

Darwin had the temperament and the knowledge of a primatologist, but the discipline of primatology itself did not develop as a professional, organized body of knowledge until the early twentieth century, when the psychologist Robert Yerkes began to experiment with captive apes after World War I. Soon it became clear that monkeys were more numerous and less costly to acquire and maintain than apes. During the decades that followed Yerkes's early experiments, Western behavioral and biomedical scientists imported thousands of monkeys, especially rhesus macaques from Asia, various monkeys from South America, and baboons from Africa. When laboratory populations increased, surplus animals were occasionally released onto small islands and other unpopulated areas, where something like fieldwork could be practiced. Even though these populations did not occupy the habitat in which they had evolved, scientists could still observe behaviors that were not being deliberately manipulated in an experimental setting.[8]

At the midpoint of the twentieth century, with improved transportation and communication infrastructures, true fieldwork became feasible. The annals of field primatology began to include stories about charismatic animals native to some of the most challenging and remote areas on earth, written by some of the most iconic figures in modern science. Thus, the evolution of the field narrative as a genre reflects the development of the discipline of primatology, as well as the changing conditions in natural primate habitat, which is increasingly under siege from human encroachment.

Some of these scientists write about their work and their study animals in terms of heroic individualism. Others write stories about themselves as participant observers in fluid, complex societies in which individual animals are moving parts in a larger whole. The genre of popular primatology

field narratives written by scientists originated with the publication of George Schaller's *The Year of the Gorilla*, an account
of his yearlong sojourn in Central Africa in 1959. He and Jane
Goodall, who began her study of chimpanzees in Central
Africa in 1960, wrote tales of romance and adventure. Dian
Fossey, who started working with mountain gorillas about
ten years after Schaller's year with them, published a field
narrative entitled *Gorillas in the Mist* in 1983—and, at least in
artistic and media representations, lived a tragedy. Biruté
Galdikas embarked on a quest into the wilds of Borneo in the
early 1980s and wrote about her work with orangutans in a
spiritual autobiography, *Reflections of Eden*, published in 1995.
Like Goodall and Fossey, Galdikas presents her animal subjects as complex, larger-than-life characters.

   Although field scientists were able to study monkeys in
the wild and publish significant scientific studies about them
by the middle of the twentieth century, popular field narratives about monkeys did not appear until Schaller and Goodall had already established the genre with their ape stories.
Sarah Blaffer Hrdy first published her book about colobine
monkeys, *The Langurs of Abu*, in 1977; it contains features of
the popular primatology narrative, although educated general
readers were evidently not Hrdy's original intended audience.
On the other hand, *Almost Human*, Shirley Strum's 1987
account of her baboon fieldwork, was clearly written for the
lay reader as well as the scientist.

   These books by Hrdy and Strum resemble novels, as they
are defined by one of the most influential literary theorists of
the twentieth century, Mikhail Bakhtin. In Bakhtin's view, the
novel is a fluid narrative genre, composed of multiple story
layers, referencing multiple literary forms, hospitable to multiple voices, and, above all, emphasizing the subjectivity and
psychological autonomy of not only the author but also other
personages in the story.[9] The primatology field narratives I
discuss in this book have these qualities; although most of

them can be read as novels, the works by Strum, Hrdy, and Sapolsky resemble the messy, "dialogic," and fluid narrative that Bakhtin theorizes as most "novelistic." Unlike the more technical scientific literature, these informal accounts convey rich, imaginative pictures of the interior lives of primates going about their daily business in their own worlds. They also convey what it feels like to be a close observer of those worlds, sometimes to the extent that the felt realities of the study animals blur with the author's own.

The Langurs of Abu and *Almost Human* also bear the feminist imprint of their time. The feminism in these books is a function not only of the political climate of the late twentieth century but also of what philosopher of science Thomas S. Kuhn has famously called a "paradigm shift," or a shift in basic beliefs and values that allows for a new wave of discovery.[10] Field scientists had much to teach those whose work was carried out in the laboratory. Following Goodall's protocols, many field scientists came to believe that animals could be better understood if they were known as individuals, with names instead of numbers. As social scientists moved away from behaviorism—which, of course, focuses on quantifiable behaviors rather than the subjects' interior lives—primatologists became more willing to grant their animals increased agency and something like consciousness. As a result of this shift, scientists can now speak with much greater confidence about animal behavior and cognition. They are also more willing to speculate about consciousness and agency.

An emphasis on consciousness and agency in study animals is just a trend, of course, and there are some notable exceptions to it. For example, in two highly readable books—*How Monkeys See the World* (1990) and *Baboon Metaphysics* (2007)—Dorothy Cheney and Robert Seyfarth narrate years of observations and field experiments with vervets and baboons, concluding that nonhuman primates in general do not have a complete theory of mind. In other words, Cheney and Seyfarth do not believe that monkeys and nonhuman apes fully know *what* they know, or fully understand *how* information is available (or not) to others. Cheney and Seyfarth pull back at the paradigm shift toward linking animal behavior with the possibility of animal interiority. Nevertheless, on the whole, narratives about primates have become more interesting—indeed, more literary—since the paradigm shift that occurred after Jane Goodall's entry into primatology. Even though their animal characters are not finely drawn, Cheney and Seyfarth do introduce narrative elements to charm and hold the interest of readers accustomed to the vivid stories told by others in their discipline. If these two scientists do not participate fully in the trend of these informal field narratives, they are still influenced by it.

And the field narrative genre as *story* remains vital into the twenty-first century. Robert Sapolsky—younger and more jaded than Strum or Hrdy—loved his monkeys as much as anyone, but, striking out for new literary territory, he wrote *A Primate's Memoir* (2001) as a self-deprecating parody of the field study. Sapolsky's work with baboons in Kenya was originally conceived as a supplement to his controlled research in the laboratory. Certainly, Sapolsky's background in experimental science explains his anxious oscillation between seeing the baboons as subjects and seeing them as objects. But his anxiety also seems to result from a feeling, shared by many of his colleagues and explained in his book of essays, *Monkeyluv* (2005), that primatology has become a female-dominated discipline. Statistics say otherwise, but the attitude about female dominance has perhaps had an impact on research, and women certainly predominate in popular media representations of wild primates and the scientists who study them.

In *A Primate's Memoir*, Sapolsky ruminates on human nature as much as baboon nature—and underscores the similarities. Sapolsky's difficulties in the laboratory and the field are intensified by pressures on fieldwork in primate habitat, as humans interfere more often and more destructively in the lives of the animals. Indeed, since primate habitat is giving way to human encroachment, field narratives as adventure stories are becoming more difficult to write. The field itself— as natural primate habitat—is almost gone, or drastically changed.

At this moment, the primatology field narrative is still a living genre, and, as Vanessa Woods's *Bonobo Handshake* (2010) demonstrates, new discoveries are still being made about primates in locations where they evolved. But the planet is in danger from deforestation, pollution, human overpopula- tion, loss of biodiversity and ecological balance, and climate change. Most primate habitat happens to be in environmental hot spots—that is, locations where these problems are expressed most dramatically. Unless the current economic and geopo- litical trends that destroy primate habitat are reversed, most nonhuman primates may soon exist only in zoos, laboratories, and reserves. If that happens, field narratives such as those I discuss in this book will be replaced by other kinds of stories. As wild populations go, so goes the literary genre about them. The reverse may also be true.

Some environmentalists complain that it is easier to inspire concern for monkeys (and other charismatic mega- fauna) than for monkey habitat. It would be nice if humans were more farsighted. However, as Darwin suggested, sympa- thy for a being like oneself, which is written into human evo- lutionary history, may be the root of ethics and morality. So if apes, monkeys, and prosimians are saved in landscapes where they evolved simply because humans somehow identify with them, then the forests and forest fragments that serve as the lungs of the planet will be preserved; the fresh water that is its lifeblood will be in greater supply; and global warming can perhaps be reduced or stalled.

Any story that inspires action based on sympathy and understanding is, in practical terms, a good story. The rich narratives of field primatology have such potential. If recent literary and scientific theorists are to be believed, the telling of stories is encoded in human DNA, and, since primates resemble one another so much, it makes perfect sense that those who spend their time with our next of kin would write stories about them, in the same way that we humans tell stories about members of our own species. These stories are an important part of both modern science and contemporary literature.

The stories I examine here are only a sample of the available primatology narratives (indeed, many of them exist only within the extremely popular genre of wildlife documentary). I have chosen these books because they suggest how primatologists adapt existing literary forms to convey their particular experiences, which are as varied as the primates they study. Not surprisingly, the generic development of primatology narratives roughly parallels the development of narrative forms in Western history, from the classical to the postmodern. These story forms seem so inevitable that one is tempted to say that humans only had to develop speech in order to tell them, and they make powerful rhetorical statements by engaging and challenging their readers.

In *Sense of Place and Sense of Planet*, Ursula Heise makes the point that certain genres can often "override" the stories that "fit less well into existing narrative patterns"—in other words, uncomfortable stories.[11] The particular genius of these storytellers is their ability to deploy "comfortable" genres to do uncomfortable work. The stakes could not be higher: stories can reveal and also shape the world, and stories about our fellow primates can contribute to saving them from extinction. But the disappearance of species and habitats constitutes only one kind of loss. It is sad to remember that there will be no more stories by Shakespeare, Charlotte Brontë, or James Baldwin, all of whom made their audiences uncomfortable.

Primatology field narratives have influenced the ways in which humans understand animals, but the stories change as habitat and field conditions change, and the impact of the stories is likely to wane as the genre changes; it may be a vicious circle. We humans require challenging stories. We consume them avidly in multiple forms and at all times. As long as there are apes, monkeys, and prosimians, there will be stories about them. But if these animals are confined in zoos, sanctuaries, and even small managed reserves because their species survival necessitates human intervention and manipulation, what a loss that will be for those of us who crave stories about animals, love, death, politics, and the wild!

CHAPTER ONE

First Contacts

There is something, some essence of Darwinism,
which is present in the head of every individual
who understands the theory.

—RICHARD DAWKINS, *The Selfish Gene*

I

Stories about apes and monkeys reveal our deepest hopes and
fears, and for the last century and a half, the figure of Charles
Darwin has brought these hopes and fears into focus. Both
his theories and his numerous anecdotes about primates of
all kinds illustrate the deep kinship between humans and our
order mates.

To understand why Darwin's attention to primates has
been a flash point and an inspiration, it is helpful to begin a
little before his time, with a glance at one of the last philosophers of the Enlightenment period, Immanuel Kant. Kant
was no animal lover, and his work does not concern apes or
monkeys, but it has a bearing on what those who came after
him have thought about them. In 1781, with the *Critique of
Pure Reason*, Kant challenged an Enlightenment article of
faith: that since reason accounted for the success of human

civilization, by fostering this God-given faculty, humans could be perfectible as social, political, and moral beings. Since the human infant's mind was what another Enlightenment thinker, John Locke, termed "a blank slate," education assumed central importance in the construction of human identity. Education, it was posited, could strengthen and train reason, one individual at a time, in order to bring about a more advanced civilization. You are what you're educated to be, according to this branch of Enlightenment thought. Not so fast, said Kant: the infant mind is not a blank slate; humans are actually born with mental frameworks, or "categories," such as an innate understanding of time and space, which determine in large part what we can learn, how we learn it, and how we deploy our knowledge later on. Furthermore, Kant added a few years later in the *Critique of Judgment*, every individual is entitled to a certain amount of irreducible "subjectivity" or uniqueness in perception and taste—another aspect of mind that lies outside or beyond reason.

These are hopeful thoughts if one values individual creativity, but they can also be terrifying, for they leave the door open for the crazies, monsters, and beasts that philosophers such as Locke had almost shut out from the definition of what it is to be human. Edgar Allan Poe, for instance, saw the terrifying aspects of the new wave in Enlightenment philosophy. Kicking and screaming, he held on by his fingernails to the notion of reason. He wanted to believe in universal order and rationality, but he was afraid that disorder and emotion ruled human affairs. *Think* what will happen, he reiterated, if we let the beast in! What if the beast is already in? What if we can't control it? Enter Poe's progenitor of the fictional detective hero, the Parisian Auguste Dupin—aristocratic, refined, reclusive, abstemious, apparently devoid of all sexual feelings, and rational almost to the point of being a disembodied brain. In "The Murders in the Rue Morgue," Poe's first Dupin story, written in 1841, brain meets body, reason meets brute, and reason wins, but with a nasty surplus of nagging, unresolved suggestions about human and beast.

The tale begins with a discourse on analytical reason, which, according to Poe's unnamed narrator, can be better demonstrated by playing cards than engaging in "all the elaborate frivolity of chess."[1] Although chess is complicated, it is mechanical; on the other hand, whist requires the successful player not only to calculate but also to observe and analyze the other players' expressions of emotion as well as their method of play. Poe calls this kind of analysis "ratiocination," the rational analytical approach of the successful detective, who must grasp both the rational and irrational behaviors of his fellow humans. Almost from the beginning of classical Western civilization, powerful stereotypes of apes and monkeys began to emerge. As Poe understood it, these figurative (and sometimes actual) apes and monkeys suggest interesting mirrors for human emotions and actions.

After Dupin reads in the evening newspaper about the grotesque and puzzling murder of a mother and her adult daughter in the Rue Morgue, he seizes the opportunity to demonstrate his own mental prowess, solving the case on the basis of newspaper accounts alone. One account reports that the murder occurred on the fourth story of an almost empty house; it describes the mutilated bodies, one shoved with brute force up the chimney, the other lying on the pavement below. Another news story contains interviews with witnesses on the street, who heard two loud voices—one belonging to a Frenchman and the other variously identified as Spanish, Italian, Dutch, German, or Russian. Remarkably, every witness is from a different European nation, and each identifies the second voice as speaking a language he does not understand. That being the case, Dupin remarks to his friend, "You will say that it might have been the voice of an Asiatic—of an African. Neither Asiatics nor Africans abound in Paris; but, without denying the inference, I will merely call your attention to three points. The voice is termed by one witness 'harsh rather than shrill.' It is represented by two others to have been 'quick and *unequal.*' No words—no sound resembling words—were by any witness mentioned as distinguishable."[2] Therefore, the detective concludes, the voice is not a human voice: this is the crucial insight that enables him to solve the case without viewing the crime scene. Although Dupin has the solution right away, he decides to visit the scene in order to develop a strategy for flushing out the murderer and proving his conclusion.

According to police reports, the door and windows to the apartment were locked from the inside at the time of the murder, but Dupin discovers a hidden egress from a window and measures the considerable distance between the window and a lightning rod that would have provided the only way down. Just as the voice had been bizarre, "there was something *excessively outré*—something altogether irreconcilable with our common notions of human action"—about these killings: the height the perpetrator had to descend in making an escape, the superhuman strength required to shove the daughter's body up the chimney, the decapitation of the mother with a few strokes of a straight razor, the strange disarray of heavy furniture, and the abandonment of money bags in plain sight.[3]

In short, the "murderer" is a Bornean orangutan, a species about which Dupin has read in the works of Baron Georges Frederic Cuvier, a founder of comparative anatomy. It turns out that the human voice was that of a French sailor, who was able to negotiate the tenuous handholds on the side of the building almost as easily as the ape because he has had years of experience in the rigging. Operating on this conjecture, Dupin places a false ad in the newspaper, luring the sailor to his apartment, where, at gunpoint, the frightened man confesses that he owns the orangutan. The ape has killed not from malice but from the instinct to imitate his master, who was shaving when the orangutan snatched the razor and escaped with it. The disarray in the apartment and the strange disposal of the bodies are the result of the beast's "consciousness of having deserved punishment" and his efforts to conceal his misdeeds.[4]

The game is over and reason returns: Dupin has solved the case, an innocent suspect is released, the prefect of police makes a partial and grudging concession, and the orangutan is sold for a large sum to the zoo. But loose ends remain, story particles that are indeclinable, irreducible. The setting of the murders in the Rue Morgue suggests a link between the murders in question and the murders of other unidentified victims of unidentified crimes by other unidentified murderers. The orangutan's voice makes sounds that are identified by "earwitnesses" as human language. Even though these witnesses are mistaken about the speaker, they have made a perfectly understandable error. In more modern times, we are able to distinguish between a human voice and a machine-generated voice on the telephone on the basis of intonation, and humans with cybernetic voice-producing implants, such as those given to laryngeal cancer patients, inevitably sound different from other people. Intonation, then, is a distinctive feature of human language, and the same is true of animal communication. Thus, Poe drives home the similarities between human beings and other animals.

If Dupin has restored reason within the frame of the story, Poe was unable to do so outside the frame. Kant's work had compromised faith in the limitless power of reason, and only a few decades before Poe wrote his story, the French Revolution had temporarily destroyed this faith by enshrining reason and committing atrocities in its name. At the same time, science was beginning to suggest that evolutionary kinship with apes was a real possibility—although, in the industrialized nations, fear of other primates developed long before Darwin's work formulated and confirmed the close kinship between humans and nonhuman primates. Poe lived in frightening times.

However, Poe's source for information about apes, Georges Cuvier, tried to alleviate the fear of the atavistic human. The most respected scientist in France during the early decades of the nineteenth century, Cuvier carried out foundational work in paleontology, comparative anatomy, and zoology. As a paleontologist, he sifted through the fossil record offered by the countryside around Paris and developed a theory of periodic catastrophes, in which all life ended, followed by successive creations. So, unlike many other European scientists who struggled with fossil evidence for a chain of life, Cuvier assured his contemporaries that humans were not at the end of an unbroken evolutionary chain that might have included ape ancestors. As an anatomist, Cuvier specialized in skeletal comparisons, and he concluded that differences in anatomy determined differences in function—not the other way around. Thus, humans were physiologically fitted from the get-go for technological prowess, and apes for an arboreal life (though it must be added that, in Cuvier's time, so little was known of the anthropoid apes that even their nomenclature was contested). At least to the satisfaction of many of his scientific contemporaries, Cuvier effectively sealed off the human from the ape by using comparative anatomy to reinforce his theory of separate creations.

One particularly nasty aspect of Cuvier's comparative anatomy was its emphasis on differences among races, which, he speculated, might have been created separately, as separate species or varieties with separate physical and mental abilities. Indeed, many scientists at the time found the geographical proximity of the darker races to the pongid apes suggestive. Cuvier's work was a boon to proslavery Europeans and Americans, just when the forces of humanitarianism and democracy were threatening slavery, a labor system that had resulted in enormous wealth for the industrialized world. How convenient to speculate that the different races of humans might also be different species!

Such racist implications of Cuvier's work can be found just below the surface of Poe's story, in which uncivilized Asians and undercivilized French lower classes are potentially savages. Here, Poe's narrator muses, after perusing a page from Cuvier's work, "It was a minute anatomical and generally descriptive account of the large fulvous Ourang-Outang of the East Indian Islands. The gigantic stature, the prodigious strength and activity, the wild ferocity, and the imitative propensities of these mammalia are sufficiently well known to all. I understood the full horrors of the murder at once."[5] But the murders are not murders because they are not committed by a human being in command of his faculties. What, then, of other murders, committed by beings without full control of their faculties—individuals of other races, other classes, other ways of life less rational than one's own? Are these crimes really murders if the perpetrators' actions are determined by different internal categories, beyond the reach of reason? Poe's story celebrates reason, but with an undercurrent of fear that there is more to the human experience than rationality and cognition. For Poe, this "more" was not simply, as Kant would have it, the benign categories of time and space, or a morally neutral preference for raspberries over strawberries or the music of Rameau over Mozart, but an indefinable horror.

Poe was not the first to document the conscious or unconscious fear that our bodily likeness to other primates disposes us to lose ourselves beyond the boundaries of reason, and he was certainly not the first to suggest that some humans are closer to animals than others. But "The Murders in the Rue Morgue" was written just when these issues were gravitating to the raging center of scientific and cultural debates about species, on the one hand, and human rights, on the other. Racial differences were exaggerated, the taxonomy of simians was debated, the line between human and animal species was often blurred, and even scientists tended to be defensive about biological and possibly psychological similarities between humans and apes.

So "The Murders in the Rue Morgue" serves as a good starting place to examine the powerful images of nonhuman primates in Western culture. The apes and Africans in the Tarzan stories of Edgar Rice Burroughs, the monkeys and native Indian villagers in Kipling's *Jungle Books*, and the fierce tribesmen and monstrous gorilla in *King Kong* have both created and reinforced fears of the simian within. Although this book will focus on Western attitudes, the possibility that apes embody our own brutal selves is not just a Western idea. For the moment, one example, I hope, will suffice. Thomas Savage, an early explorer and natural scientist working in west Africa, made the following observation, quoted by Thomas Henry Huxley in *Man's Place in Nature*: "It is a tradition with the natives generally here, that [gorillas] were once members of their own tribe: that for their depraved habits they were expelled from all human society, and, that through an obstinate indulgence of their vile propensities, they have degenerated into their present state and organization."[6] Although these beliefs about nonhuman primates are not a cultural universal, they have been widely shared by humans all over the world—and swept into a vicious circle of racism, anthropocentrism, imperial exploitation, abuse of monkeys and apes, and destruction of their habitats.

II

The slipperiness of taxonomy—how the physical bodies of various life-forms are described, distinguished, and classified—contributes to political and scientific confusion in discussions about primates, human and otherwise. The eighteenth-century Swedish biologist Charles Linnaeus revolutionized biology by creating a logical classification system that accounted for all known species and allowed for the systematic naming of newly discovered species. Partly as a result of Linnaeus's work, fossil collecting was all the rage in nineteenth-century Europe, and as paleontologists worked their way through successive geological strata, they discovered a fossil record of changes in species over geological time—though some, like Cuvier, managed to find a way to dismiss the possibility of speciation.

In any case, Charles Darwin was not the first to entertain the idea of evolution. Of the numerous pre-Darwinian attempts to account for apparent evolutionary changes in life-forms, the *Philosophie zoologique*, published in 1809 by Cuvier's rival Jean Baptiste Lamarck, is the best known. Lamarck suggests that organisms develop traits in order to take better advantage of their surroundings: over time, orangutans would have developed long, strong arms and short legs in order to thrive in the jungle treetops, from which they rarely descend. The teleological perspective of Lamarck's theory is still tempting today, and even scientists sometimes fall into Lamarckian language. Certainly, an individual orangutan uses its long arms in order to swing from tree to tree or branch to branch, but orangutans as a species did not develop those arms for the purpose of brachiation. Instead, as Darwin would later have it, by imperceptible gradations, it happened that longer-armed orangutans were more successful at getting food where they lived, avoiding enemies, and finding mates in their jungle environment. Hence, they passed on these traits to more offspring until, over the course of innumerable generations, orangutans evolved very long arms. Although many found Lamarck's account of biological transmutation compelling, it remained controversial for fifty years before Darwin's competing story of evolution took center stage.

In 1858, in great excitement, Darwin's acquaintance Alfred Russel Wallace sent him an essay on evolution that accurately summarized the ideas Darwin had been working out privately for twenty years. Darwin had been developing his theories ever since an 1831–36 voyage of scientific discovery on the *Beagle*, which famously culminated with the young man's astonished observations of the unique wildlife of the Galapagos Islands. (His adventures became a best-selling travel narrative right away.) Although he was reluctant to publish before building the strongest possible case for his theory of evolution, Darwin quickly decided to present the theory before the Linnean Society in London, giving due credit to Wallace. The theory was out and Darwin put forth a Herculean effort, so that by the following year *The Origin of Species* appeared in print. This work established Darwin's scientific trajectory from then on and inspired a controversy that continues to this day.

Certainly, the idea of evolution was "in the air" at midcentury, and it has been a historical truism to say that Darwin's success was due to the timeliness rather than the originality of his discoveries. But Darwin's case for evolution became dominant because it was supported by profound scholarship, attention to the best scientific methods known at the time, and voluminous evidence from fossils and live specimens (his membership in an intellectual elite also helped by opening access to these specimens). Darwin asserted that minute variations in a species are "selected" by nature if they are advantageous to individuals in ways that allow them to reproduce more successfully than their fellows. Thus, the new trait is passed on, and over eons a species may accumulate so many of these variations that it transmutes into a new species. A variety or species that does not change will likely die out because the conditions of life inevitably do change.

Darwin refers to a special form of natural selection as "sexual selection," a process whereby individuals select as reproductive partners other individuals who indicate special fitness, whether through strength, weapons, or ornaments. These attributes enable some individuals (usually males) to win contests for mates (usually females) by appealing to them rather than just relying on brute force. Evolutionary biologists since Darwin have been fascinated by this insight and pursued it from the female's point of view: a boy lightning bug who flashes more often gets the girl, and the peacock with the most dramatic plumes gets the peahen, in both cases because the female surmises that the extravagant male really does have more stamina and therefore better survival chances and better genes. In Darwin's scheme, developed at much greater length in *The Descent of Man*, sexual selection is one aspect of natural selection. Furthermore, such choices might, Darwin suggests, indicate that humans are not the only animals to have an aesthetic sense. However that may be, multiply these natural and sexual selection processes by the millions, and the result is an almost unfathomably long history during which myriads of species arise and then pass away.

A current myth about Darwin is that he delayed publication of *The Origin of Species* until 1859 because of his personal struggle between the scientific evidence he saw with his own eyes and a commitment to Christian teachings. In his youth, Darwin did prepare for the ministry because he knew that country living would afford him the leisure and setting to pursue his studies in natural history (especially collecting insects). But in matters of religious faith, he had already tended toward agnosticism in the 1830s, while sorting out the implications of the impressive collections he had made, the life-forms he had observed, and the geology he had studied during his five years on the *Beagle*. After marrying Emma Wedgwood in 1839, Darwin became a dedicated family man, and he regretted differing from his wife on religious matters. But he delayed publishing his findings because he wanted credibility within the tough and increasingly professional scientific community as much as he dreaded giving offense against religious orthodoxy.

Thus, when he published *The Origin of Species* the year after presenting his and Wallace's findings before the Linnean Society, Darwin said next to nothing about our fellow simians and left the history of human evolution to be filled in by attentive readers. Darwin was an animal lover, and he was especially fascinated by apes and monkeys, but he wanted his theory to be taken seriously and saw no reason to roil the waters by placing *Homo sapiens* in the middle of the argument and openly claiming kin with other primates. On the other hand, Darwin's friend and tenacious defender Thomas Henry Huxley had no such reservations. To this day, many of those who are nervous about evolutionary theory speak of a "Darwinian conspiracy," a perception that can be attributed to Huxley as the organizer and center of the X Club, which was dedicated to publicizing and popularizing Darwin's theories and discrediting those who disagreed, especially on religious grounds.

Huxley was a righteous bulldog. In 1863, four years after the appearance of *The Origin of Species* and somewhat to Darwin's dismay, he published *Man's Place in Nature*, a series of ethnological lectures he had given at the Royal School of Mines. This little volume left nothing about our place in the primate family tree to the imagination. The frontispiece was the now-famous illustration of a *Homo sapiens* skeleton walking tall at the head of an evolutionary line, followed by various anthropoid skeletons, with a skeletal gibbon last in line. Critical as it was of previous ethnological, paleontological, and biological scientists, the appearance of Huxley's little book forced the pace of professionalization in the sciences and brought to light the full implications of Darwin's theory: that humans, too, are animals, in the same family tree as the apes and with no pretensions to a separate creation. According to Huxley, "The question of questions for mankind—the problem which underlies all others, and is more deeply interesting than any other—is the ascertainment of the place which

Man occupies in nature and of his relations to the universe of things. Whence our race has come; what are the limits of our power over nature, and of nature's power over us; to what goal we are tending; are the problems which present themselves anew and with undiminished interest to every man born into the world."[7] Huxley's pugnacious attitude was especially useful in terms of primatology, in its rudimentary stages an extremely contested discipline, as it is still today. Indeed, primatology is a conflictive disciplinary zone probably for the reason Huxley himself gives: that our relation to nature is the fundamental question and that investigations of nonhuman primates more powerfully (and more viscerally) suggest answers than the pursuit of any other knowledge.

Not only was the taxonomic relationship between humans and other primates contested, but even basic data about apes and monkeys was difficult to collect and unreliable. The great apes, especially, occupied territories almost impossible for Europeans to reach, and the apes feared humans as much as humans feared them, with better reason. Like other field studies during the eighteenth and nineteenth centuries, primatology usually consisted of tracking, shooting, dissection, preservation in kegs of alcohol, and skeletal reconstruction— or killing adult apes in order to capture infants, who usually died in adolescence even if they survived the ocean voyage back to a European zoo. There were a few exceptions among naturalists, but it is worth remembering that, in order to paint them, even John James Audubon killed, mounted, and stuffed the birds he loved. In spite of the tireless collecting done during this period, primate taxonomy remained difficult because the transportation of large specimens, such as the cadavers of great apes, was so challenging that few whole specimens were available. Behavioral studies were next to impossible: few naturalistic primate groupings could be observed in zoos, so data about intraspecies social behavior had to come from field studies, and field biologists had to contend against great obstacles, without well-defined scientific protocols. Scientists relied on whatever lore they could dredge up from missionaries and explorers, their own fragmentary observations of apes fleeing for their lives, or stories they heard from the locals, many of which turned out to be tales for children! Tall tales mistaken for true accounts, misidentifications of specimens, and profound disagreements were inevitable. And, in fact, the taxonomic record is still under construction as new species of monkeys continue to be found and known species are reclassified after closer study.

In the first section of his book, Huxley did his best to sort through the available information about apes from Ovid onward, distinguishing myth from fact, explaining the sources of confusion, composing clear and well-organized descriptions of anthropoids, and suggesting future directions for scientists. In the second chapter, Huxley argues that if humans are more like gorillas than gorillas are like gibbons, then we must accept our place in the ape family tree. He follows this thesis with every comparison he can make among the specimens available: he takes skeletal measurements; counts and categorizes teeth; describes nostrils, eyes, hands, and feet in detail; records information on fetal development; and observes configurations of preserved human and anthropoid brains. Finally, Huxley calculates brain weights by filling braincases with millet seed, substituting and weighing an equal volume of water, and, since brain tissue weighs 10 percent more than water, multiplying the water weight by 1.1.

For Huxley, the evidence for claiming kin to other primates is overwhelming. He recommends a thought experiment, in which the reader might imagine the puzzlement of a space explorer from Mars who attempts to classify earth species and understand how humans can logically justify placing themselves in their own category. (Darwin would return to this idea in *The Descent of Man*, citing Huxley and pointing out that "if man had not been his own classifier, he would never have thought of founding a separate order for his own reception.")[8] Not only does Huxley insist that humans are apes; since evidence for biological distinctions among races is inconsequential in comparison to the rest of the data he has collected for his study, he posits an early version of what we now call the cultural construction of race, arguing that racial divisions are based on flimsy or nonexistent science. Cuvier might have contributed a few facts to the study of fossils and primate skeletons—and to the detective stories of Poe—but he was not one of Huxley's heroes.

Now the war between religious conservatives and scientific professionals began in earnest, the battles so bitter that they still loom large in our cultural awareness. In England, the famous 1860 debate on evolution between Huxley and Bishop Samuel Wilberforce (who accused Huxley of having apes for ancestors) was a source of hilarity in the press and hard feelings for the contestants. Sixty-five years later, in the United States, the "Monkey Trial" of the Tennessee high school biology teacher John Scopes was so notorious that it has affected public perception of scientific education in the United States ever since. And if the U.K. has finally put Darwin on its money, in the United States, only about half the general population believes that Darwin had it right, and school districts remain torn about the appropriateness of teaching "Darwinism" without a religious counterweight, currently called "intelligent design." Today, even among those who take Darwin seriously, the extent to which Darwinism can be applied to human psychology and society continues to be a topic of sharp debate. Both social Darwinism (which cannot justifiably be called "Darwinism" at all) and evolutionary psychology are still anathema to many social scientists and humanists.[9]

In any case, after Huxley was seconded by other prominent scientists of the day (notably the feisty German Ernst Haeckel, whose search for fossil evidence of "the missing link" was even more provocative than the projects of British evolutionists), Darwin finally had to take charge of the argument about our place in the great tree of life.[10] As usual, his concern was to preserve not only the reality but also the appearance of scientific objectivity—a challenging task under the circumstances. *The Descent of Man*, published as two heavy volumes in 1871, begins with barely concealed exasperation: "During many years I collected notes on the origin or descent of man without any intention of publishing on the subject, but rather with the determination not to publish, as I thought that I should thus only add to the prejudices against my views."[11] This book provided the same kind of scholarly underpinning to the theory about human origins and our kinship with other simians that *The Origin of Species* had supplied for the general theory of evolution and speciation. True to form, Darwin had more to say than he anticipated. The two best-selling volumes of *The Descent of Man* were not enough, and the following year he added a third, *The Expression of the Emotions in Man and Animals*, which was even more appealing to the public because readers could test Darwin's conclusions against their own observations and the generous illustrations.

Once committed to the struggle, Darwin drew upon all his resources—the notes he had collected for years; the texts he read voraciously; the specimens now pouring into museums and private collections from the edges of the expanding British Empire; a voluminous correspondence with scientists, explorers, missionaries, and nabobs all over the world; and conversations with impresarios of animals shows and, especially, zookeepers. In a series of somewhat domesticated adventures, Darwin spent as much time as he could at zoos, which at that time were repositories for both preserved specimens and live animals.

In *Primate Visions*, Donna Haraway describes Jane Goodall's account of her adventures in Tanzania as a "first contact" narrative.[12] In a way, Darwin's zoo excursions might also be considered first contacts, since he undertook them in order to understand kinship rather than estrangement—in other words, to see these primates as embodied beings, not as they had been presented in myths and tall tales. Since he had donated many of his own collections to the London Zoo, Darwin enjoyed the cooperation of administrators and keepers in conducting numerous experiments. He presented various yarns to weaver birds to test their color preferences for nest building and tried to make elephants weep. He provoked monkeys to test their reactions; he gave a doll to one individual to gauge its surprise and snuff to another to see if it would close its eyes when it sneezed (it did not). He presented a mirror to orangutan adolescents, watching them caper and pose before scampering away in alarm.

Darwin's first encounter with an ape, in fact, had been in 1838 at the London Zoo. Jenny, an infant orangutan who wore a dress and lived in the heated giraffe house, impressed him with her human-like emotions, her understanding, and her ladylike deportment (she had been presented to the Duchess of Cambridge). Contrary to the descriptions of Cuvier and his predecessors, whose assessments of primate aggression were weighted with value judgments, Darwin noted in a letter to his sister Susan that when Jenny's keeper teased her by showing her an apple and then taking it away,

she threw herself on her back, kicked & cried, precisely like a naughty child.—She then looked very sulky & after two or three fits of pashion, the keeper said, 'Jenny if you will stop bawling & be a good girl, I will give you the apple.[']—She certainly understood every word of his, &, though like a child, she had great work to stop whining, she at last succeeded, & then got the apple, with which she jumped into an arm chair & began eating it, with the most contented countenance imaginable.[13]

Orangutans would come under Darwin's scrutiny again almost twenty years later, when he read four articles on the apes by Wallace in the *Annual Magazine of Natural History*, published soon after Wallace's return from the Dutch East Indies. Darwin's younger colleague had observed adult orangutans in the wild, and once, after his companions shot a mother orangutan, he kept the baby in his camp. She eventually died from malnutrition, but not before impressing him with her intelligence and her emotional similarity to human children, including a tendency toward tantrums. (These accounts by Darwin and Wallace coincide almost exactly with Galdikas's descriptions of infant orangutan tantrums, affection, dependency, and intelligence.)

In his later volumes, Darwin seized the opportunity to argue for sexual selection, which his colleagues had greeted with skepticism and he had thus minimized in *The Origin of Species*. In Darwin's view, the evolution of any species, including humans, depends largely on sexual selection. Although his original genetic research had to be conducted with rapidly reproducing nonhuman species (especially mollusks and pigeons), in the *Origin of Species*, Darwin explained the mechanics of sexual selection with evidence spanning the whole animal kingdom, including crustaceans, insects, fish, birds, reptiles, and amphibians. The focus of *The Descent of Man* is the primate order: in arguing for human kinship with other animal species, Darwin relies heavily on fossil evidence of archaic humans and secondary data about other primate species. In the first chapter, by repeating Huxley's arguments, adding a few details, and placing the whole discussion within the context of speciation, Darwin gives his stamp of approval to the theory that humans, like other species, are simply the most recent stage in an evolutionary chain that—granting a few missing links—goes back to an ape-like ancestor.

However, whereas Huxley's morphological argument ends with an assertion of bodily likeness between humans and other apes, Darwin's explanation begins with morphology and quickly moves on to arguing similarities in human and animal cognition, psychology, society, and culture. In fact, Darwin distrusted Linnaean classification, and in *The Descent of Man*, he notes especially the case of the New World capuchin monkeys, the various forms of which some naturalists rank as species and others as varieties: "If of a cautious disposition, [the classifier] will end by uniting all the forms which graduate into each other as a single species; for he will say to himself that he has no right to give names to objects which he cannot define."[14] For Darwin, the study of life should concentrate not on external form but on processes within the lives of individuals and over evolutionary time. Darwin's arguments about animals, including humans, are informed by attention to behavior and even attempts to understand motivation. He reasoned that human emotions and human society provide clues for understanding the lives of other animals, especially nonhuman primates, who can in turn serve as a mirror for a deeper understanding of human behavior.

In Darwin's view, then, it matters less how many millet seeds it takes to fill up a brain case than how the brain— or any other organ—works. In *The Descent of Man*, Darwin begins with taxonomy and morphology, but most of his additions to Huxley's work on nonhuman primates are physiological rather than morphological details. In his discussion of skin and hair, for instance, Darwin suggests that the relative scarcity of hair on human bodies evolved partly as a result of sexual selection and perhaps also because relative hairlessness helps humans remain free of parasites (a function of social grooming among other primates).

One of Darwin's sources of information for *The Descent of Man* was a study of New World monkeys made by a German physician and explorer of Paraguay, Johann Rudolph Rengger, whose interests, like Darwin's, lay in function and behavior over form. Against the trend of the time, and in spite of obstacles, Rengger made keen observations about the behavior of capuchin monkeys and speculated about their physiological processes. Citing Rengger, Darwin argues that close ties between humans and other primates are strongly suggested by these monkeys' susceptibility to human diseases—bad colds, tuberculosis, "apoplexy, inflammation of the bowels, and cataract in the eye. The younger ones when shedding their milk-teeth often died from fever."[15] More interesting than these maladies is the monkeys'

strong taste for tea, coffee, and spirituous liquors: they will also, as I have myself seen, smoke tobacco with pleasure. Brehm asserts that the natives of north-eastern Africa catch the wild baboons by exposing vessels with strong beer, by which they are made drunk. He has seen some of these animals, which he kept in confinement, in this state; and he gives a laughable account of their behaviour and strange grimaces. On the following morning they were very cross and dismal; they held their aching heads with both hands and wore a most pitiable expression: when beer or wine was offered them, they turned away with disgust, but relished the juice of lemons. An American monkey, an Ateles [spider monkey], after getting drunk on brandy, would never touch it again, and thus was wiser than many men. These trifling facts prove how similar the nerves of taste must be in monkeys and man, and how similarly their whole nervous system is affected.[16]

Today, diseases transmitted from humans are one of the leading causes of captive primate death, and even a casual visitor to research and shelter facilities is sometimes required to provide documentation of a tuberculosis test. Darwin was clearly intrigued by Rengger's research—and both men would perhaps be puzzled by how often biomedical and behavioral researchers continue to test the simian nervous system by giving monkeys cigarettes, beer, and other addictive substances.

Other likenesses Darwin notes between the human senso-rium and that of other primates include similarities in nose, eyes, ears, and vocal apparatus. Humans have more promi-nent noses than most other primates, but it is impossible to distinguish humans from all other primate species on this basis, since the Hoolock gibbon has an aquiline nose and some monkeys, such as the proboscis monkey, carry the nose "to a ridiculous extreme."[17] All primates have similar facial structures; Darwin does not mention binocular or trichro-matic vision (his resources for investigating neurology were limited), but he does note that many nonhuman primates have human-like eyebrows. Like some humans, certain mon-keys also have vestigial Mr. Spock points on their ears, and though a distant ancestor common to all primates probably had moveable ears, this trait has not been preserved in apes: "The ears of the chimpanzee and orang are curiously like those of man," Darwin writes, "and I am assured by the keepers in the Zoological Gardens that these animals never move or erect them; so that they are in an equally rudimentary con-dition, as far as function is concerned, as in man."[18] (Some humans can wiggle their ears, however; it's a good party trick.) Darwin goes on to speculate that the strength and arboreal habits of the great apes and archaic humans might have pro-tected them from the kinds of danger that would have made sharper hearing necessary for ground-dwelling species.

Of all primate physiological functions, Darwin seems most intrigued by the vocal prowess of some monkeys and apes, particularly the "singing gibbon," whose long calls are more powerful than any human voice, even that of a trained opera singer. Gibbon calls, remarkably, travel farther even than the long calls of gorillas and orangutans. One gibbon, the *Hylobates agiles*, Darwin notes, can reproduce an entire octave of musical notes, "which we may reasonably suspect serves as a sexual charm," just as music functions in the human species.[19] (Curiously, W. C. L. Martin, one of Darwin's sources, reproduces the call in musical notation.) Many simians, Darwin remarks, are musical and convey emotions by tonality, as Poe anticipates when his witnesses to the murders in the Rue Morgue comment on the orangutan's supersyllabics and tone. Darwin thus believes that most primate calls function as communications of important information about the individual's surroundings or emotional state, or they serve as a kind of vocal ornamentation in reproductive competition among males. But among the New World species studied by Rengger, both sexes of some howlers evidently sing because "they delight in their own music and try to excel each other."[20] In this way, Darwin speculates, monkey calls are probably similar to the music and poetry of archaic humans. Ever intrigued by sex, he proposes that in humans singing evolved before speaking—because it added sex appeal and could express emotion as well as information.

Darwin also devotes a great deal of space to the limbs and extremities, but unlike Huxley, he explains function along with anatomy. Darwin's investigations lead him to believe that evolutionary pressures among the various primate species have resulted in hands specialized for climbing (monkeys and apes) or refined manual dexterity (humans), but it is physically impossible, he theorizes, for a hand to be perfectly suited for both. Likewise, Darwin considers legs, feet, and toes as they have adapted to locomotion, concluding that bipedalism, an adaptation to open spaces, would be maladaptive in arboreal environments, where food, shelter, and safety depend on ease of movement in trees. He suggests that human bipedalism initiated a cascade of other adaptations—feet that support weight, a spine curved for balance, a broadened pelvis. Bipedalism also freed the hands, making powerful jaws and large teeth unnecessary as weapons and even less important for chewing, since food could be manually modified.

But Darwin stops short of relating increased intelligence and capacity for speech to bipedalism, as some more recent evolutionary biologists have suggested. In his view, human mental powers originated in primate social life in all its dimensions. Indeed, almost everything Darwin has to say about primate morphology, bodily functions, and senses is connected to speculations about behaviors shared by humans and their cousins, the apes and monkeys.

Like most twentieth-century primatologists, Darwin believed that the behaviors most clearly connected with physiology are reproductive. In his view, sex is so central to male primate experience that nonhuman male primates demonstrate an ability to distinguish females of other species from the males. In the second edition of *The Descent of Man*, Darwin added an account of a male mandrill at the zoo who attempted to bully women visitors into mating. Evidently figuring his best chances for reproductive success, "he was by no means aroused with so great heat by all. Always he chose the younger and picked them out in the crowd and summoned them by voice and gesture."[21] This anecdote evidently so embarrassed Darwin that he not only placed it in a footnote but wrote it in Latin; presumably, women, who were seldom trained in classical languages, would not be able to read it. (Without denying this particular mandrill's behavior, I must point out that recent studies show that nonhuman male primate sexual behavior is more often a response to female readiness and that older females are generally more popular than young ones, who have yet to prove their fecundity.)

In animals that reproduce through internal fertilization, of course, males and females have different organs. But Darwin almost always subordinates his discussions of morphology to speculations about process and function. In most primate species, adult males tend to be larger boned, louder, bigger, hairier, and more colorful than adult females. A South African Little Red Riding Hood might have said "What big teeth you have!" to a strapping male chacma baboon instead of a wolf. The question is why. Darwin suggests that sexual dimorphism is apparently unrelated to family structure, whether monogamy or polygamy, and he observes that both kinds of families have occurred in human communities throughout history. (However, the relationship between sexual dimorphism and social reproductive structures is still debated among primatologists today.) What clearly does matter in Darwin's scheme is sexual selection—that is, a differential in the ability of individuals to pass on their genes. One determiner of sexual selection is male competition for mates; size and strength are useful in a pitched battle. The size and strength that help an individual find mates also help when males take on the role of defending their female and infant conspecifics from predators—a behavior related to reproductive success insofar as it protects a particular male's progeny or the progeny of his kin.

Maybe big, strong males are more attractive to females even without a fight, especially if they are handsome and musical. Darwin goes into detail about what makes the males of various species sexy. Male primates tend to have more powerful vocal mechanisms, and in most primates with throat pouches, the males' are bigger than the females'—if the females have them at all. Darwin interprets this feature as an advantage in battles among males for sexual access; as a sexual attraction to females; and, in some species, as an aid to the male in fulfilling his responsibility to protect the females of his group and their young. Likewise, Darwin interprets the large canines and sagittal crests, or cranial reinforcements, of gorillas as defensive advantages that come into play most often during competition for, or protection of, mates. He explains the beards of human males as sexual ornamentation and the relative hairlessness of human females in terms of sexual attractiveness. Similarly, in a long and charmingly illustrated discussion of sexual dimorphism in monkeys, Darwin interprets various puffs, crests, points, and contrasting hair colors in male capuchin, spider, and langur monkeys as sexual ornamentation, useful in attracting females, who are in many primate species the chooser rather than the chosen. Darwin notes the similarity between simians and humans: "Man is more powerful in body and mind than woman, and in the savage state he keeps her in a far more abject state of bondage than does the male of any other animal; therefore it is not surprising that he should have gained the power of selection."[22] This belief accounts for the circular logic implied by the "pornographic" Latin footnote and critiqued by feminists of Darwin's own time and place: women are not as intelligent as men and therefore should not be trained to read Latin; then, their inability to read Latin suggests a lower degree of intelligence. Darwin was a revolutionary, but he was also a man of his time.

Although Darwin's science went against the grain of many Victorian prejudices, it was so much a product of the time that his argument on sexual dimorphism in humans is not only anthropocentric and Eurocentric but more convoluted than most of his other work. In this section of his analysis, which covers not geological time but historical time and geographic location, modern civilized women are said to have "evolved" ornamentation by stealing bird plumes and other artificial body extensions to attract men. Humans are therefore different from other primates in the operation of sexual selection. According to Darwin's theory of sexual selection, in the rest of the primate world (with a few exceptions), males are larger than females because they must fight one another to vanquish rivals and impress the females, who then choose to mate with the victors. The relatively large size of human males cannot be explained in these terms, he claims, because men no longer fight over women. In spite of his diligent research into the reproductive habits of other primate species, Darwin is reduced to special pleading: human males have inherited their relatively large size "from some early male progenitor, who, like the existing anthropoid apes, was thus characterized."[23]

Darwin was a poetry lover in his youth and packed Milton in his bag before boarding the *Beagle*. Nevertheless, he seems to have overlooked the significance of the Victorian fad of inventing and reinventing an epic past in literature, art, architecture, and even home decorating, with fictional heraldry placed above every bourgeois fireplace and suits of armor decorating the entrance hall. One such reinvention was Alfred Tennyson's Arthurian epic *Idylls of the King*, published serially between 1859 and 1872, easily the most popular poetic work of the day. Book 10, "The Last Tournament," was coincidentally published in the same year as *The Descent of Man*. Like many Victorian works of poetry and song, *Idylls of the King* is about the chivalric virtues to which Victorians aspired, but almost every plot point involves fighting over women. Men continue to fight over women and bully them into submission even in the twenty-first century, and while most of these battles are conducted with weapons of words or music (Mick Jagger) or money (Donald Trump), that is unfortunately not always the case. To be perfectly fair to Darwin, though, it must be admitted that the complex connections among physiology, behavior, and reproductive success have continued to be one of the most contentious themes in modern primatology, not to mention biology, anthropology, ethnology, sociology, and psychology.

Darwin continued to revise *The Origin of Species* throughout his life, integrating new information, closing gaps, and answering friendly and unfriendly critics. His theories of natural selection and the struggle for existence—which became known in common parlance as "evolution" and "survival of the fittest"— were soon generally understood and applauded among most of the scientific community.

III

So, in Darwin's scheme of things, an investigation of species origins necessitates an investigation of sexuality in particular and behavior in general, as well as—one discovers in following Darwin's train of thought through all of his books—investigations of mind and emotion, not just in humans, not just in primates and other animals, but sometimes even in plants.

It might surprise (or appall) those critics of Darwin who have not studied his work that almost a third of the first volume of *The Descent of Man* is an impassioned analysis of connections among emotions, mental abilities, and morality in humans and their mammalian kin. Although courtship, mating, and embryonic development are strikingly similar in all mammals, in infancy and adolescence, humans are even more obviously related to apes and monkeys than to other mammals: Darwin points out that monkeys are nearly helpless when they are born and that most orangutans are not mature until they are ten or more years old—not very different from humans in some societies. Consequently, parenting, which in humans is usually institutionalized as motherhood, is essential to the survival of the species. In one of the warmest passages in *The Descent of Man*, Darwin recounts anecdotes of human-like motherly devotion: a capuchin driving flies away from her infant; a gibbon washing the faces of her children in a stream; captive primate mothers whose grief at losing an infant is so intense that they are suicidal; the adoption of orphan monkeys by unrelated adult females; monkey mothers carefully dividing food among the young; female baboons mentoring the young of other species; even a female baboon who adopts a kitten. (In the first edition of *The Descent of Man*, Darwin comments on the intelligence displayed by this baboon, who bit off the kitten's claws after she was scratched. After a critic expressed doubt about the episode, Darwin added a footnote in the second edition, claiming that he bit off a kitten's claws himself to prove that it could be done by primate teeth.) Although some of these claims have not been verified in modern behavioral observations, they are still good evidence of Darwin's views. In his opinion, then, the primate maternal instinct is so strong that it can operate independently of parturition or genetic kinship. Maternal instincts are allied to and perhaps the origin of other forms of altruism, including, Darwin suggests, the moral instincts developed by archaic humans.

Following his discussion of the maternal instinct, Darwin develops a more general description of the social instincts. "I fully subscribe to the judgment of those writers who maintain that of all the differences between man and the lower animals, the moral sense or conscience is by far the most important," he remarks.[24] But he also observes that "any animal whatever, endowed with well-marked social instincts, would inevitably acquire a moral sense or conscience, as soon as its intellectual powers had become as well developed, or nearly as well developed, as in man."[25] In addition to the nurturing activities that can be associated with mothering, Darwin argues that warning cries, social grooming, mutual defense, cooperative hunting, obedience to a leader, and defense of the weak by the strong can be interpreted as protomoral behaviors. Even vengeance and jealousy, if not moral in themselves, reveal a mental capacity that can be channeled into moral behavior, as Darwin illustrates through an anecdote added in the second edition:

Sir Andrew Smith, a zoologist whose scrupulous accuracy was known to many persons, told me the following story of which he was himself an eye-witness; at the Cape of Good Hope an officer had often plagued a certain baboon, and the animal, seeing him approaching one Sunday for parade, poured water into a hole and hastily made some thick mud, which he skilfully dashed over the officer as he passed by, to the amusement of many bystanders. For long afterwards the baboon rejoiced and triumphed whenever he saw his victim.[26]

Two additional anecdotes, first given in volume 1 and later repeated in the "Summary and Concluding Remarks" at the end of volume 2, exemplify Darwin's thinking about the moral or protomoral instinct in primates. In the first anecdote, he writes,

I will give [an] instance of sympathetic and heroic conduct in a little American monkey. Several years ago a keeper at the Zoological Gardens, showed me some deep and scarcely healed wounds on the nape of his neck, inflicted on him whilst kneeling on the floor by a fierce baboon. The little American monkey, who was a warm friend of this keeper, lived in the same large compartment, and was dreadfully afraid of the great baboon. Nevertheless, as soon as he saw his friend the keeper in peril, he rushed to the rescue, and by screams and bites so distracted the baboon that the man was able to escape, after running great risk, as the surgeon who attended him thought, of his life.[27]

A baboon is the hero of the second anecdote, from Alfred Edmund Brehm, like Rengger one of Darwin's most reliable and complete sources of information on primate behavior. Brehm observed that when a troop of baboons fled from a pack of dogs, the dominant males managed to hustle all of the adolescents out of the way except one, who got stranded on a boulder. When one old patriarch, "a true hero," returned to the boulder, "coaxed him, and triumphantly led him away," the dogs were so "astonished" that they did not attack.[28] Darwin notes that many mammals, including and especially dogs, have instincts for the "more complex emotions" of loyalty and fellow feeling, but the focus in this part of his argument remains on primates.[29]

Since there were no behavioral science labs in Victorian England, the only people able to give detailed accounts of animal cognition were zookeepers and the impresarios of animal shows and menageries. In fact, since their livelihood depended on intelligent, reliable animals, impresarios and animal trainers developed strategies not only for training their animals but also for identifying those with the most potential. One of Darwin's sources was a Mr. Bartlett, whose approach was based on his observations of attention span: the longer a monkey was able to pay attention without being distracted, the greater its potential as an actor, and the more Mr. Bartlett was willing to pay for the animal. Since, in Darwin's analysis, moral or protomoral behavior is allied to intelligence, the discussion of morality in *The Descent of Man* is bound up with the discussion of primate intelligence, which (in a parallel to Kant's explanation of the human mind) is a function of both inherited capacities and learning. For René Descartes, whose 1637 treatise *Discourse on Method* initiated the European cultural emphasis on reason, humans were the apex of organic life, qualitatively different from the lower animals in their capacity for reason, which was expressed in human language and only human language. In contrast to the human, animals were always on automatic pilot, according to Descartes and the mainstream of Enlightenment philosophy that developed from his work. Although a century later the Utilitarian philosopher Jeremy Bentham pointed out that this analysis of animal nature did not excuse the mistreatment of any creature that could feel pain, the idea of "organic machines" was the default position in science and philosophy until Darwin returned to it.

Darwin saw traces of reason in the animals he studied, just as he saw traces of pure instinct in human nature. In one of the most interesting passages of his discussion of nonhuman mental powers, he quotes Alexander von Humboldt's *Personal Narrative*: "The muleteers in S. America say, 'I will not give you the mule whose step is easiest, but *la mas racional*,—the one that reasons best'; and Humboldt adds, 'this popular expression, dictated by long experience, combats the system of animated machines, better perhaps than all the arguments of speculative philosophy.'"[30] If mules are smart, then nonhuman primates are smarter. A century before Jane Goodall's revelations about the lives of chimpanzees, Darwin found in his wide reading and research evidence of apes and monkeys using tools, building nests, passing on "culinary" skills such as the best way to eat an egg, and teaching botanical knowledge to the young.

Darwin was not a primatologist but a generalist, and all the evidence he draws from the nonhuman primate world supports his contention that humans are descended from an ape-like ancestor, through the mechanisms of the struggle for existence, natural selection, and sexual selection, which work just the same in human evolution as they do in the evolution of dogs or deer mice, cucumbers or cockatoos. He concludes the analysis of human descent by noting that embryology, physiological vestiges, and homological structures point to one conclusion about our ancestry:

that man is descended from a hairy quadruped, furnished with a tail and pointed ears, probably arboreal in its habits, and an inhabitant of the Old World. This creature, if its whole structure had been examined by a naturalist, would have been classed amongst the Quadrumana, as surely as would the common and still more ancient progenitor of the Old and New World monkeys. The Quadrumana and all the higher mammals are probably derived from an ancient marsupial animal, and this through a long line of diversified forms, either from some reptile-like or amphibian-like creature, and this again from some fish-like animal.[31]

Further back, our ancestors were gilled hermaphrodites, and further still, we were similar to the larvae of ascidians—

simple, sack-shaped water critters such as the sea squirt. But in Darwin's view, physiology made no sense apart from the mental and emotional attributes that humans share with other mammals; morphological and physiological evolution could not be separated from the evolution of behavior.

Although he was horrified by slavery and by the brutal decimation of Indian tribes that he witnessed on his travels in South America, Darwin was not a "liberal" in the sense that we now use the term. On the voyage of the *Beagle* thirty-five years before the publication of *The Descent of Man*, he had studied the human inhabitants of Tierra del Fuego. Like Thomas Malthus, who contemplated the poor of the industrialized European nations, Darwin found the existence of the Fuegans "nasty, brutish, and short." He preferred claiming kin with the animals, and, recalling his friend Huxley's encounter with Bishop Wilberforce, he comments at the very end of *Descent* that "for my own part I would as soon be descended from that heroic little monkey, who braved his dreaded enemy in order to save the life of his keeper; or from that old baboon, who, descending from the mountains, carried away in triumph his young comrade from a crowd of astonished dogs—as from a savage who delights to torture his enemies ... and is haunted by the grossest superstitions."[32] Darwin had so much to say about mind, emotion, culture, and society that he had to write another volume, published a year after *Descent* as *The Expression of the Emotions in Man and Animals*. According to Darwin's biographers Adrian Desmond and James Moore, this smaller book amounted to "the amputated head of the *Descent* that had assumed a life of its own."[33] It was a best seller; readers could test every assertion, comparing each of the many illustrations to the faces of people they knew, to the body language of their pets, and to the monkeys they saw in the zoo. Unlike the history of life on earth, which had to be extrapolated by experts from geological strata and an imperfect fossil record, expressions of emotion were evident for all to see.

No less an authority than Konrad Lorenz, often considered the originator of the discipline of ethology, credits Darwin with originating ethology, for it was Darwin who articulated the notion that "behavior patterns are just as . . . reliably characters of species as are the forms of bones, teeth, or any other bodily structures."[34] However, while Darwin himself was willing to set aside the Enlightenment dictum that mules and monkeys were animated machines, most twentieth-century scientists found it difficult to give up the view that animal minds are simply bundles of inherited or conditioned impulses without true consciousness. And most scientists today, trained in the language of behaviorism, continue to guard carefully against the appearance of the sin of anthropomorphism. Many have felt that attributing an inner life to nonhuman animals is, in itself, anthropomorphic.

It may be "scientific" to avoid speaking of other primates as if we share vast tracts of psychological experience, but avoiding "anthropomorthic" language is difficult, because most people feel that we are psychologically kin to other primates as well as morphologically similar. In English, even the language used to describe groups of apes and monkeys reflects this difficulty. There are parliaments of owls, prides of lions, and gaggles of geese, but (at least in common parlance) troops of monkeys—as of soldiers, actors, or circus performers. Snakes, weasels, and cockroaches are "it," but most English speakers readily grant grammatical gender when speaking of simians; they are "he" and "she." Cats have kittens and dogs have puppies, but our fellow primates have babies, sons, daughters, grandmothers, and so on. The English language contains no other words for genetic relationships in primate families. Although popular accounts of primatologists' lives and the lives of their study subjects admit social and psychological kinship with other primates, the technical literature persists in endeavoring to appear objective by censoring this kind of language. I have noticed this difference even when comparing the published abstracts of oral presentations at scientific primatology meetings with the oral presentations themselves. In the abstract, the typical presenter rigorously avoids humanizing language—even sometimes grammatical gender. The result is stiff and occasionally awkward phrasing. But in the oral delivery, that same presenter's language is natural, humanized, and even humorous, though sometimes apologetically so.

Darwin had no such reservations. His conclusions about the expression of emotion are based on careful and detailed studies of facial anatomy and physiological investigations of many species. Contrary to the contemporary practice of primatology in the West, he cultivated anthropomorphism—in animal studies, we would now call it "critical anthropomorphism." Although one may be skeptical of animal intelligence, and although one must set aside entirely any speculations about the spirituality of animals, Darwin argued, it is beyond question that humans share with them emotions and mutually understandable expressions. In *The Expression of the Emotions in Man and Animals*, Darwin recycled the field studies he had used in his previous works, but he relied even more on direct observation. He studied his dogs and the family cats, his favorite horse Tommy, farm fowl and other neighborhood birds, the cattle in nearby fields, and, in the same spirit, the antics and expressions of his children. He went back to the London Zoo, where he spoke with the primate keepers charged with the care of chimpanzees, orangutans, gibbons, baboons, Barbary apes, macaques, and several species of New World monkeys. "Some of the expressive actions of monkeys," he explains, "are interesting . . . from being closely analogous to those of man."[35]

In the monkeys and apes Darwin was able to observe, he saw little difference between the expression of affection and that of pleasure and joy. Monkeys laugh and smile when they are pleased, and Darwin notes that when chimps and orangutans are pleased, their eyes sparkle—especially when they are tickled. Like humans, monkeys also crinkle their eyelids when happy or amused. Attention, games, food treats, and reconciliation after quarrels can all be sources of pleasure for these primates. Some primate expressions of joy, such as grinning and baring the teeth, can be confused with expressions of pain or anger, but with a little practice, an observer can almost always distinguish subtle differences.

Not surprisingly, Darwin notes that nonhuman primates are equally expressive when they are in pain or otherwise unhappy. All of them express pain or unhappiness by crying, and a few even shed tears, as humans do. As far as Darwin was able to observe, other primates don't frown, but they do express rage as emphatically as any human being—by striking the ground with a fist, yawning, screaming, staring, pouting, raising their eyebrows, reddening with passion, sucking their teeth, and throwing tantrums. Pain and astonishment could be read clearly on the countenances of the zoo animals Darwin studied. He recounts the almost comic cognitive dissonance of a multispecies group of monkeys who encountered a turtle in their cage for the first time. Curiosity enticed them to come close, and some of them stood up for a better view before reconsidering and running away. Some made jabbering noises, which Darwin interprets as attempts to conciliate the turtle. On other occasions, he observed monkeys raising their eyebrows in wonder before tasting a new food or when listening to a strange sound. When terrified, some monkeys raise their eyebrows, some scream, and some lose control of their bowels. Sometimes, the hair of a frightened monkey stands on end. And once Darwin watched a monkey "almost faint from an excess of terror" when caught.[36]

Humans are equipped to interpret most animal emotions, Darwin concludes at the end of a long section on animals, because, like our own feelings, animal emotions can be explained according to three principles. The first is the principle of "serviceable associated habits," which have evolved in mammals, especially, as surely as protective coloration or canine teeth for hunting, because the outward expression derives from a physiological process. An open mouth denoting astonishment, for instance, is the outward expression of a sudden intake of breath, which might be needed for a quick escape.[37] The second principle is "antithesis": in some situations, there is an involuntary tendency to express an opposite emotion, such as smiling in a conciliatory way to mask fear.[38] The third Darwin calls the principle of "direct action of the nervous system," such as trembling in fear.[39] Darwin is clearly fascinated by the expression of emotions in different kinds of animals, who are treated in discrete sections of the book and constantly brought into the more general discussions of anatomy and physiology. But the point of this study is that humans and other mammals express emotions *in the same ways and for the same reasons*. So much for animated machines.

These were Darwin's lifelong convictions, not so much drawn from his research as motivating it. In a private notebook, he remarked to himself, "Animals—whom we have made our slaves we do not like to consider our equals.—Do not slave holders wish to make the black man other kind? Animals with affections, imitation, fear. pain. sorrow for the dead."[40]

IV

It would be almost impossible to exaggerate Darwin's influence on science and society. His work in what are now an array of separate scientific disciplines has been formative. The effect of his famous books on the scientific literature of his time, not to mention fiction, poetry, and drama, was substantial. Darwin's contributions to the modernist worldview are almost incalculable; the figure of Darwin continues to be omnipresent not only in science but in cultural forms of all kinds.

Aside from his theories of natural selection, sexual selection, and the struggle for existence, Darwin's greatest contributions to science and culture have been to help us identify the important questions and to inspire debate—even when this means mounting challenges to the great scientist himself. Built on Darwinian foundations but departing from them are developments in evolutionary biology such as cladistics, the neutral allele theory, and the notion of punctuated equilibrium. Mendelian genetics quickly proved Darwin wrong about the actual mechanics of genetic inheritance, and a full understanding of genetics was not achieved until the discoveries of Watson, Crick, and Franklin. Darwin's attitudes about gender and race were considered, in his own day, humane, although in hindsight one finds them mixed. He was every bit a creature of his own moment in history.

Darwin's work is neither perfect nor definitive, but it was revolutionary and foundational. And like all successful scientific theories, Darwin's theory of natural selection has remained authoritative because, in spite of gaps, it still explains more phenomena than any competing theory and rests on a profound underpinning of research and experimentation. Darwin was not only a methodical and diligent scientist; he was a creative thinker. In *The Selfish Gene*, Richard Dawkins suggests that Darwin's theory is so multifarious and flexible that "each individual has his own way of interpreting Darwin's ideas." Yet Darwin's ideas are also so compelling that "there is something, some essence of Darwinism, which is present in the head of every individual who understands the theory."[41] Like so many other post-Darwinian scientists, Dawkins consciously stands on Darwin's shoulders to critique and extend

Darwinian thought, within the boundaries of biology and beyond.

Darwin trumps Cuvier. Had he written "The Murders in the Rue Morgue" just thirty years later, Poe would have been hard-pressed to premise his detective story on a science that taught the separate creation of humans and beasts, black people and white. Darwin changed the course of Western cultural and intellectual history, as a theorist, storyteller, and personal example.

CHAPTER TWO

The Primatology Romance

Why take the style of those heroic times?
For nature brings not back the mastodon,
Nor we those times; and why should any man
Remodel models?

—ALFRED, LORD TENNYSON, "The Epic"

I

Like Darwin, Jane Goodall set out to do one thing and did something else. When Darwin came home after his five-year voyage on the *Beagle*, he expected to settle down and make sense of his experience and his data, while living the quiet life of the country gentleman scientist. Instead, he was swept up in a storm of his own making. Like Darwin, Goodall expected a simple life; she wanted to live with the animals in the wilds of Africa and understand them, never dreaming that her reentry into industrialized society would be fraught with intellectual controversy and laden with ethical significance. Darwin and Goodall occupy unique places in history: these two scientists have made profound contributions to the sum of knowledge and the methods for adding to it. They cannot be imitated, and their worlds cannot be restored. Their significance goes far beyond their particular discoveries, and everyone who comes after them is, in a sense, belated.

Modern primatology owes Darwin a debt, not only for key concepts and foundational theories but also for the long continuance of the Victorian craze for fossil hunting—which urged Goodall's mentor Louis Leakey forward into primatology—and for Darwin's contributions to the ways in which the stories of science are told. In chapter 1, I outlined some of Darwin's contributions to scientific storytelling, especially when the subject was primates. Here, I will additionally suggest that the figure of Darwin himself, as a hero of science, was a powerful model, directly and indirectly, for scientists in the twentieth century. From the seventeenth century forward, the heroic man of science was not an unfamiliar trope in Western culture. But it was from Darwin, in particular, that Goodall inherited the model of the scientist as questing hero. True, Darwin's example came to her along a circuitous route, but she was profoundly influenced by it nonetheless, as she, in her turn, has transmitted this model to others in the field.

II

"Meme" is a word invented by Richard Dawkins to indicate a cultural particle that replicates itself in human consciousness.[1] Dawkins conceives of genes as self-serving, and memes, in his view, behave the same way. In evolutionary terms, children's culture is a perfect medium for memetic replication, a primordial soup in which many kinds of memes replicate, sometimes at prodigious rates, or lie dormant for centuries. Like other memes, those carried in children's literature often jump back and forth through the membrane between stories and real-life actions. The human who talks to animals is a meme that originated in the deeps of antiquity and today is familiar to everyone who has read even a few books written or rewritten for children. In her 1999 memoir *Reason for Hope: A Spiritual Journey*, Goodall proposes a direct link between the influence of this meme in her own childhood reading and the development of her later career, which has included both science and storytelling: "As a child I was not at all keen on going to school. I dreamed about nature, animals, and the magic of far-off wild and remote places. Our house was filled with bookshelves and the books spilled out onto the floor. When it was wet and cold, I would curl up in a chair by the fire and lose myself in other worlds. My very favorite books at the time were *The Story of Doctor Dolittle*, *The Jungle Book*, and the marvelous Edgar Rice Burroughs Tarzan books."[2] Goodall herself contributed copy for the back cover of *The Story of Doctor Dolittle*, which was first published in 1920 and reissued in 1988: "Any child who is not given the opportunity to make the acquaintance of this rotund, kindly, and enthusiastic doctor/naturalist and all of his animal friends will miss out on something important. Start with the first in the series, *The Story of Doctor Dolittle*, and you will not be content until the others are lined up on your bookshelves. If only there were more." (There are, in fact, fifteen.) It is telling that, in the acknowledgments for her monumental 1986 study *The Chimpanzees of Gombe*, Goodall again mentions Hugh Lofting's stories, which she says inspired her at the age of eight to go to Africa to be with the apes and monkeys.

Goodall and the children's literature that shaped her can be conceptualized as carriers of a very old and powerful meme complex. In addition to her important discoveries, it is perhaps Goodall's participation in the quest romance memeplex that seems to strike such a powerful chord in contemporary culture. Nonhuman animals need a hero, and in this time of awful environmental challenges, so do the rest of us. Goodall is not simply an adventurous scientist but also a model of kindness to and respect for individual animals. That interest in individuals and belief in individuality in nonhuman species determined Goodall's deliberate reinvention of primatological field protocol and has remained a key component of her thinking during the second part of her long career, as she has shifted from science and natural history to environmental activism and animal advocacy.

Lofting's Doctor Dolittle is a naturalist, teacher, and leader who saves humans and animals from disease, ignorance, and natural disasters. In this fictional character, the figure of Darwin as a hero of science intersects with the tradition of the questing hero of romance. And it is not surprising that Lofting's stories prefigure the present-day cultural phenomenon of Jane Goodall, not only as she presents herself in her writing but also as she has appeared in various media representations: a tall, blond, youthful woman with a ponytail, dressed in khaki and sturdy shoes, making her way through the forest and touching, following, or exchanging glances with the animals she studies and serves (or more recently, in a lab coat, comforting caged primates used for biomedical research). So powerful is this new recombinant meme (recombinant partly because the hero of primatology is a woman) that, in popular primatology and sometimes even the scholarly literature, many scientists who have come along since Goodall have had to position themselves as followers and imitators, or to deliberately and explicitly distance themselves from this model.

To reconstruct the Darwin-Dolittle-Goodall hero genealogy, it is useful to glance back briefly at the history of the quest romance, one of the oldest literary forms in Western culture. The best-known quest romances are those of King Arthur and the knights of the Round Table, characters who came into cultural prominence a little before 1200, when their exploits were described in Geoffrey of Monmouth's *History of the Kings of Britain*. These characters and others like them dominated the popular imagination in Western Europe for about four centuries. In her history of the English quest romance, Helen Cooper argues that after the romance plays of Shakespeare, such as *The Tempest* (1610), the meme of the lonely questing hero dropped out of mainstream literature for a while, but it did not die out. It was simply transferred into a different alembic—folk culture and stories for children. It also appeared in an occasional religious allegory quest on the model of John Bunyan's *The Pilgrim's Progress* (1678) and, in the last century or a little more, science fiction.[3]

The most obvious feature of the genre is the hero. He is young and untested; like the fool in the Tarot deck, who sets out with a tiny pack, a little dog, a high heart, and unwary steps, the romance hero at the beginning of his journey is innocent and hopeful. The quest plot has four parts. First is the call to leave the known for the unknown. The middle of the story is a series of adventures as the hero makes his way through threatening forests or trackless wasteland. In some of these adventures, the hero succeeds in accomplishing good or winning a fight, but in some he does not. There are obstacles and dangers of many kinds—impassable thickets, evil fairies and wicked warriors, voracious beasts or monsters, illusory or enchanted castles. But the hero also encounters helpers along the way—Merlin, the great magician of Arthur's court, for example, as well as hermits, kindly peasants, fellow knights, good fairies, special animals such as dogs and horses, or friendly wild or mythical creatures. Although the goal of the quest may emerge or evolve as the hero meets helpers, obstacles, and adventures, what is clear from the beginning is the hero's willingness to take risks in order to find the Holy Grail, rescue the princess from the tower, or obtain some other object of desire that arises from the journey itself. Despite the expectation of a happy ending, at the climax of the tale, the hero comes within sight of the goal and may or may not succeed in accomplishing what he set out to do. Either way, the quest romance ends with the hero's coming home to a place that has changed, for better or worse, and into which he must find or make his place anew. Thus, the hero's adventures involve not only physical struggle but also what Cooper terms "self-becoming," and in the end his interior development is at least as important as his contests with evil magicians, hostile knights, deceptive fairies, and wild beasts.[4] The hero may pass these tests or fail them, but he will discover that part of the goal was the testing itself.

Darwin was a different kind of hero—the Victorian scientist and man of authority—but if he is compared to the meme of the romance hero, there are similarities enough to have inspired Hugh Lofting; his hero Doctor Dolittle bears more than a passing resemblance to Darwin. Darwin engaged in both a literal journey of adventure and a dangerous lifelong intellectual quest, to reach goals that emerged slowly.[5] His first published work, *The Voyage of the* Beagle, was a best seller almost fifteen years before he was known as a controversial theorist. Although in this book Darwin did broach tentative versions of what would be his most controversial claims later on, it was the narrative of his adventures and observations on land and sea, along an exotic route of places of interest for imperial Britain, that excited the imagination of the British reading public.

Doctor Dolittle is a naturalist of the early Darwinian stamp—"the greatest naturalist in the world," his young helper Tommy Stubbins insists. The doctor collects specimens during long, dangerous voyages and brings them home to Puddleby-on-the-Marsh for closer study. Darwin's specimens were usually dead, in contrast to Dolittle's living (and voluntary) zoo, but this is a children's story, after all. Darwin dissected his specimens and studied fossils; Dolittle's knowledge comes from talking with the animals in his collection, a skill he learns from his parrot, Polynesia, who in turn learns from the doctor that her imitations of human speech actually have meaning. (Irene Pepperberg, the trainer of the African gray parrot Alex, would confirm the literal truth of Polynesia's learning curve.) Similarly, as we have seen in *The Expression of the Emotions in Man and Animals*, Darwin claims to have learned a great deal not just from his specimens but also from his dog, the livestock on neighboring farms, and various animals he observed in the zoo. He might not have talked to the animals, but he certainly listened to them. Thus, both the real scientist and the fictional children's hero began their quests with open minds and a thirst for knowledge about the natural world. Following their hearts, these heroes reach the Holy Grail of a more profound understanding of nature and human nature within it, but what they learn does not coincide exactly with what they expected to learn.

Lofting gives an obvious nod to Darwin in his second novel, *The Voyages of Doctor Dolittle* (1923). When Tommy Stubbins asks Dolittle to teach him to read and write so that he can become a naturalist, the doctor replies, "It is nice, I admit, to be able to read and write. But naturalists are not all alike, you know. For example: this young fellow Charles Darwin that people are talking about so much now—he's a Cambridge graduate—reads and writes very well."[6] On the other hand, the South American Long Arrow does not, and he is a brilliant naturalist too, according to the doctor. There is in this volume a further, more arcane reference to Darwin. The doctor's aim in his second quest is to understand the language of mollusks. In his four monographs on barnacles, Darwin worked through many of the problems he would later tackle in *The Origin of Species*; these monographs still constitute the definitive study of the *Cirripedia*. (Of course, barnacles are not mollusks but crustaceans.)

One morning, Tommy finds the doctor in his study, a room furnished with telescopes and microscopes, cases of birds' eggs and seashells, charts of flora and fauna. The doctor is leaning over a glass box of water, listening to a "Wiff-Waff" in order to glean information about the beginning of life on the planet. "We find their shells in the rocks—turned to stone—thousands of years old," the doctor informs his young friend. "So I feel quite sure that if I could only get to talk their language, I should be able to learn a whole lot about what the world was like ages and ages and ages ago."[7] After Dolittle's failed attempts with many other mollusks and crustaceans, the Wiff-Waff seems especially promising because he is a morphological miracle: "You see he really belongs to two different families of fishes." But the creature's language is quite limited because of "the life he leads."[8] (In fact, in *The Voyage of the* Beagle, Darwin describes the South American tinochorus, which presents its own classification challenge because the species is allied to several bird genera—one among many resonances between Darwin's travelogue and Lofting's stories.) Belonging as he does to a species on the brink of extinction, the Wiff-Waff has so few companions that language would help him very little. "I have no doubt," the doctor informs Tommy, "that there are shellfish who are good talkers—not the least doubt. But the big shellfish—the biggest of them, are so hard to catch. They are only to be found in the deep parts of the sea. . . . If a man could only manage to get right down to the bottom of the sea, and live there a while, he would discover some wonderful things—things that people have never dreamed of."[9] There's nothing for it, then, but to find a ship and set sail, looking for the right equipment and the main chance.

Along the way, Doctor Dolittle, the hero of Darwinian science, morphs into the hero of the quest romance. He solves a murder by questioning a bulldog who witnessed it; saves a group of grateful bulls from being slaughtered in a bullfight in the Capa Blanca Islands; rescues his entire crew when a storm wrecks their ship; reverses the dangerous southern drift of the floating Spider Monkey Island off the coast of South America by enlisting the aid of whales; teaches the human inhabitants of the island to use fire and construct sturdy buildings; heals everyone who is sick; saves Long Arrow and his friends from entrapment in a cave after an earthquake; and finally agrees, to his own dismay, to become King Think-a-Lot, a name his subjects consider more dignified than Dolittle. In all these adventures, except that of becoming a king, one finds echoes of Darwin's *Voyage of the* Beagle. But even in Dolittle's short reign, there seems to be a veiled reference to both Darwin's humanitarian feelings and his assumptions of Anglo-Saxon superiority.

With the help of his faithful human and animal companions, Dolittle eventually manages to wriggle out of his kingly duties by healing the Great Glass Sea-snail, whose tail has been injured in the same quake that trapped Long Arrow. Once Dolittle has deciphered his language, the grateful mollusk agrees to transport the doctor, his friends, and Long Arrow's magnificent collections back to England in his transparent shell (an imperialist gesture, of course, akin to Darwin's own collecting), taking the scenic route along the bottom of the sea. Although Dolittle's quest for the beginning of life on the planet is unfinished, he learns a great deal, and all are happy to be going home. At the end of the book, Doctor Dolittle is finally able to devote his attention once more to science—until the next adventure.

If the second book in Lofting's series reveals the figure of Darwin as a cultural meme that replicated itself in the fictional character of Doctor Dolittle, it is the first book, *The Story of Doctor Dolittle* (1922), that most people know. It has inspired a play, an animated film, a filmstrip, a 1967 musical film with Rex Harrison in the leading role, and three looser film adaptations, Americanized and modernized, since 2001. (Less happily, Lofting's character has also inspired the term "Dolittle complex," which scientists have used to accuse one another of anthropomorphizing animals to the point of overestimating the communicative abilities of nonhuman animals, but that is another issue.) *The Story of Doctor Dolittle* has been adapted for younger children as well, and Bantam Doubleday has issued an expurgated version in response to the criticism that the book is racist in its treatment of Dolittle's other human sidekick, Crown Prince Bumpo, who is given to malapropisms and mindless optimism.

The accusations of racism (and imperialism) are not unfounded: the books are embedded in the cultural norms of Lofting's historical moment. However, the child Jane Goodall was obviously more interested in the cast of nonhuman characters, including apes and monkeys by the thousands, than the political implications of the human relationships in the books. One of the doctor's earliest humanitarian acts is to save the monkey Chee-Chee from a life of hardship and ridicule as an organ grinder's helper. After he is rescued, Chee-Chee takes up residence at the cottage in Puddleby-on-the-Marsh, where he has a bed in the plate rack, but he is homesick. When a swallow brings news of a plague in Africa that has infected all the primates, Chee-Chee persuades Dolittle to travel to Africa to help them. So, with his animal family, the doctor heads south to save Chee-Chee's kin.

Dolittle's ship eventually "runs into" Africa, and the crew make their way through the kingdom of Joliginki toward the Land of the Monkeys. Crucial in this endeavor is the ingenuity of the animals themselves; when the doctor and his crew come to the top of a cliff in their flight from some belligerent humans, they are saved by the monkeys and apes, who hold hands to make a bridge leading down from the escarpment. (Field observations have revealed that some monkeys and apes actually perform this feat to help infants who have not learned to brachiate.) Now the doctor can begin his work: "John Dolittle . . . became dreadfully, awfully busy. He found hundreds and thousands of monkeys sick—gorillas, orangutans, chimpanzees, dog-faced baboons, marmosets, gray monkeys, red ones—all kinds. And many had died."[10] With the aid of other jungle animals, the doctor works for "three days and three nights" to heal the sick and vaccinate the well, and in two weeks the plague is over.

The apes and monkeys are so grateful that they decide to help Dolittle with his research by capturing the renowned Pushmi-Pullyu, so that the doctor can finance his new ventures by displaying the two-headed beast to paying customers. This animal-which has since become extinct, we are told—is so rare that even the Comte de Buffon does not mention it in his vast *Natural History* (forty-four volumes published between 1749 and 1804). Dolittle asks the Pushmi-Pullyu to join them for the journey back to England, and the Pushmi-Pullyu agrees. To celebrate, the monkeys clap for a long time, and "the Grand Gorilla, who had the strength of seven horses in his hairy arms, rolled a great rock up to the head of the table and said, 'This stone for all time shall mark the spot.'"[11] To this day, remarks the narrator, monkey mothers teach their children to pause reverently when they pass the stone in the forest. Almost incidentally, on the way home, Dolittle saves his crew from the Barbary pirates, whom he converts from their evil ways.

Works in a series, as these stories are, serve as a hospitable narrative framework for a hero who never quite becomes the perfect expression of his potential. However, this hero's ever-increasing store of knowledge does change his relationship with the rest of the world and does suggest a "self-becoming." Continually tested by circumstances, Dolittle learns more about animals as species and as individuals, and as new animal friends gather around him, they help him cope ever more successfully in a world that treats them as objects for exploitation and that runs on the power of money. For the child Jane Goodall, what mattered most was kindness to animals and the well-founded hope that she could go to Africa, talk to the animals, and learn about them. After several decades in Africa, she came to resemble her model in other ways as well.

III

Merlin, the wizard of Camelot, lived backward instead of forward like ordinary human beings, and he was inhumanly wise. Jane Goodall was installed in 1960 at Gombe Stream Reserve in Tanzania by her own Merlin, Louis Leakey, who also, in his own way, lived backward. Leakey was a tireless seeker and adventurer, a brilliant publicist for anthropology, paleontology, and, late in his life, primatology. His passion, like that of many other scientists since the Enlightenment, was to rewrite the myth of human origins.

Leakey's contributions to science included prolific finds he made with his wife, Mary, his sons, and other associates and crew in Tanzania's Olduvai Gorge, part of the great geological rift running along a north-south line in East Africa. Among these finds were fossils of *Proconsul africanus*, a distinctive ape on the way to being human; the archaic, big-jawed *Zinjanthropos boisei*, or "nutcracker man"; *Homo habilis*, or "handyman," who apparently made tools; and an older ape-man, *Kenyapithecus wickeri*. The dentition of *Ramapithecus* encouraged Leakey to push the start date for humanity back, and he also relocated the cradle of humankind from Asia, where paleontologists had previously situated it, to East Africa. (With the very recent discovery of early human fossils in Dmanisi, Georgia—estimated to be 1.8 million years old— the question of when, how, and where we became human is once again more complicated.)[12] Interpreting fossils is not an exact science, however, and since no one has ever definitively settled even what it means to be human, Leakey's theories and discoveries perhaps contributed as much by provoking debate and discussion as by adding to the factual knowledge of our origins.

In any case, fossil finds and keen debates were not enough for Leakey. In the 1950s, while serving as curator of the National Museum of Natural History in Nairobi, Leakey hit upon the notion of theorizing some missing behavioral links between humans and our prehuman primate ancestors by working backward from existing species as well as forward from fossils. Although archaic human behaviors cannot be reconstructed from observations of contemporary species, understanding the scope and variety of primate behaviors can suggest many different scripts for our ancestors besides "man the hunter." The more scripts, the more explanations paleontologists have to choose from when they discover an unfamiliar bone, tool, or fossilized footprint, and the more complete the picture can become.

To this end, Leakey chose Jane Goodall to study chimpanzees, Dian Fossey to observe the mountain gorillas of the central African highlands, and Biruté Galdikas to carry out similar research on the Bornean orangutan. Goodall has summarized Leakey's contributions to the field in several publications, and her understanding of his importance was shared by the other "trimates," as these three were sometimes called.[13] Fossey indicates the significance of Leakey's contributions to her gorilla research in Gorillas in the Mist (1983), and in Reflections of Eden (1995), Galdikas goes into some detail about Leakey's contributions to her research on orangutans and the field of evolutionary primatology in general. Clearly, these women could not have conducted their research without his initiative and his tireless advocacy on their behalf, and yet each of them undertook her assigned research for reasons of her own, contributing to science and conservation in ways that Leakey did not foresee.

Jane Goodall was the first of Leakey's protégés, and her experience foreshadowed those of many other field primatologists after 1960. In 1957, she took a trip to Kenya to visit a friend, sought and found a job as Leakey's secretary, snatched every opportunity to accompany him on digs, and, in 1960, took the chance he gave her to strike out for the wilderness and live with the chimpanzees. As a young Arthur to his

Merlin, she was eager and open-minded; she was brave because she could not possibly anticipate the dangers she might encounter. Early in her first narrative, *In the Shadow of Man*, Goodall describes the moment of landing on the beach below the Gombe preserve. It is a strange passage in which she testifies to feelings that, ten years later, she still cannot quite articulate:

> *I have often wondered exactly what it was I felt as I stared at the wild country that so soon I should be roaming. Vanne [her mother] admitted afterward to have been secretly horrified by the steepness of the slopes and the impenetrable appearance of the valley forests. And David Anstey [the colonial official sent to establish Goodall at Gombe] told me several months later that he had guessed I would be packed up and gone within six weeks. I remember feeling neither excitement nor trepidation but only a curious sense of detachment. What had I, the girl standing on the government launch in her jeans, to do with the girl who in a few days would be searching those very mountains for wild chimpanzees? Yet by the time I went to sleep that night the transformation had taken place.*[14]

In all the stories about her early days at Gombe—and she rehearses this part of the plot over and over in her books—the young Jane Goodall is a hero on a quest, not for the Holy Grail of human origins sought by her mentor but for knowledge about *other* primates in the here and now.

In Leakey's view, a relatively uneducated young woman would be closer to nature and more intuitive than a man with a scientific education, perhaps more able to make contact with the apes, and more accurate in observing them because she would be free of preconceptions. Goodall conformed to the gender stereotype and simultaneously worked against it. As Leakey predicted, she demonstrated enormous patience in getting to know the chimpanzees, she was open-hearted, and she was open-minded enough to develop new protocols that allowed her to observe her subjects intimately. In order to stay in the wilderness, however, she had to become a credentialed scientist, too. After five years, Leakey sent her back to England to work toward a doctorate under the supervision of his friend Robert Hinde, a Cambridge ethologist who was interested in Goodall's holistic approach to primate research. Goodall's goals as a scientist emerged *as* she collected data, not before, and they did not fully solidify until she understood her own discoveries within the new formal theoretical frameworks she learned in her graduate work, which ended with the completion of her Ph.D. in 1966. Even though she emerged within the halls of science as an oppositional figure and has paid the price for her marginal beginnings, Goodall also eventually became the heroic "man of science"—courageous, meticulous, objective when she needed to be, and persuasive.

Leakey, Hinde, various government officials, and several funding agencies were among Goodall's most obvious and predictable helpers. But like the hero of old romance, Goodall encountered many others. Her mother, Vanne, spent the first five months with Jane in her makeshift wilderness camp and made an almost immeasurable contribution to her daughter's work. More than Jane herself, Vanne became friends with the locals by setting up an informal clinic to treat minor wounds and ailments, and she was the one who discussed the quotidian details of camp life with the local men hired to help. After Vanne left, Jane's sister, Judy, came to take photographs and keep her company for several months. To help document Goodall's work more thoroughly, Leakey then sent the wildlife photographer Hugo van Lawick, who made substantive suggestions and collected data in addition to taking photographs. Van Lawick stayed much longer than he intended to and eventually became Goodall's husband. Some of the local Tanzanian people, hostile at first because they feared that overreporting chimpanzee numbers would increase government supervision of the land, later became friends and informants; others, especially the trackers Adolf, Saulo David, Marcel, Rashidi, Soko, and Wilbert, became essential to the work. After several years, a number of local Tanzanians also began to work at Gombe as researchers.

Not surprisingly, though, a special chimpanzee, whom Goodall named David Graybeard, became her most charismatic and memorable helper. Unusually confident and sociable, this chimpanzee was the first to allow her within close range, and the other chimpanzees eventually accepted her in part because of his comfort in her presence.[15] His trust and friendship facilitated Goodall's first major discovery, when she observed him making tools to fish for termites. David also initiated, in her presence, numerous social interactions with his group mates. Like Toshisada Nishida in the nearby Mahale Mountains, Goodall started piecing together her own picture of chimpanzee social life, especially from watching David Graybeard and the matriarch Flo, whose alpha status may have given her a level of confidence that enabled her to allow Goodall's proximity. Even though Flo was prolific, her sons have been political leaders in the Gombe community out of proportion to their number. After years of friendship, Goodall watched Flo's final days and grieved when she died.

But true to the romance form underpinning Goodall's narrative, not all of the chimpanzees were helpful and open; in general, they presented her with her greatest obstacles. They were difficult to find, and like the monsters in the deep Arthurian forests, some of them could be formidably vicious. After her arrival at Gombe in 1960, four frustrating months went by before she was able to get close enough to distinguish individuals. During this time, even distant sightings were rare, so sometimes for days on end, her research was confined to examining empty sleeping nests. Once the chimpanzees allowed her to come close, most of them remained suspicious and hostile for several more months, and if her sex made her somewhat less threatening than a man might have been, it also seems to have made her an occasional target of displays. Once, a big male named Goliath shook branches above her and created a shower of raindrops, twigs, and dead leaves. On another rainy day, far away from the camp, another big male, whom she later named J. B., screamed at her, violently shook branches and trees, and finally sneaked around to wallop her on the back of her head. From a nearby tree, a mother and infant seemed to watch in wide-eyed horror. As Goodall discovered, chimpanzee attacks on other primates are not rare. In Gombe, they often kill colobus monkeys and infant baboons for meat. In her 1986 culminating study of the Gombe chimps (though not a romance, arguably her most important publication), Goodall includes the portrait of an African man who was terribly mutilated when, at six years old, he rescued his infant brother from marauding chimpanzees who were about to eat him.

Another obstacle was the topography of Gombe: numerous narrow valleys separated by high parallel ridges, running down nearly perpendicularly from an escarpment over two thousand feet above the level of the lake. Almost pathless, the land became so overgrown during the rainy season that the chimpanzees were usually hidden from view—and, at the same time, following them through the brush became next to impossible. Goodall was reminded of just how dangerous this landscape could be several years after her arrival, when Ruth Davis, a dedicated student who persisted in tracking a group of adult males into the evening, fell from a ridge and died.

The work was brutal, as Goodall notes in *In the Shadow of Man*: "My skin became hardened to the rough grasses of the valleys and my blood immune to the poison of the tsetse fly, so that I no longer swelled hugely each time I was bitten. I became increasingly surefooted on the treacherous slopes, which were equally slippery whether they were bare and eroded, crusted with charcoal, or carpeted by dry, trampled grass."[16] In addition, Goodall suffered from malaria, numerous other tropical infections, and serious sunburn. Once, she waited nervously in the underbrush while a leopard passed her by; another time, she stumbled upon an even more lethal animal, a buffalo, fortunately asleep. One evening, a deadly water cobra floated across her naked foot while she was wading near the lake shore, where crocodiles usually lurked in the daytime. The numerous unhabituated baboons who shared the shore with them were excitable and aggressive with humans from time to time, so she had to be wary of them, too. Not wanting to use her time for meals and not wanting to be burdened by carrying food, she ate little when she was away from camp and became so thin that her visiting sister was alarmed. Sometimes, she sat for hours in the rain, waiting for the chimpanzees to move. On other occasions, so that she could be near them when they woke in the morning, she slept on the ground, with nothing for breakfast except a can of beans and a cup of cold black coffee. This life of hardship and deprivation has set a pattern for many field primatologists since then: isolation, disease, danger, discomfort, and boring food are occasionally interrupted by rich discoveries—rare glimpses across the species barrier. In this book, I will note that Fossey and Galdikas, in particular, experienced similar hardships. Vanessa Woods describes something like these experiences when she tracked monkeys in Costa Rica, as late as 2007.

The "field," as it is constructed by primatologists, is located not only in space but also in time. Although the chimpanzees were not yet threatened when Goodall arrived at Gombe, the political situation in central Africa was on the brink of dramatic change. Colonies were separating from their European rulers, often by means of violent conflict. The start date of Goodall's work at Gombe was delayed as Belgian refugees from the Congo poured into the area around Gombe and Kigoma, the nearest town. So as not to waste time, she began a short study of the vervet monkeys inhabiting an island in Lake Tanganyika, but before finishing it, she was told to proceed to her destination. The armed violence erupted most problematically in 1975, when four of Goodall's graduate students were kidnapped by forty armed men who crossed the lake to Tanzania from Zaire. They were later released, but the experience was so traumatic for these young students that they made a pact never to discuss it. The threat of violence has never quite left Gombe, and, indeed, political unrest still occurs throughout much of the remaining primate habitat today.

Goodall's internal conflicts are also part of the picture. She felt responsible—and, indeed, was held responsible by colleagues in the United States—for the kidnappings. She has also always worried because, once the animals are habituated and lose their fear of human beings, they become more vulnerable to humans who would displace them, capture them, or eat them. Many years into her research at Gombe, she regretted having set up the banana feeding stations that seemed necessary for sustained observation of the family groups she wanted to study; to all appearances, the feeding stations changed the very behaviors she originally wanted to observe.

Another internal conflict developed around the conflicting demands between Goodall's approach, which was holistic and individualist, and the protocols of academic science. Leakey managed to obtain funding for Goodall's first research period and to extend that funding as she began to collect substantive data. *National Geographic* was especially generous in its support for the project, and her article "My Life Among the Wild Chimpanzees" appeared in its pages in 1963; the society later published two more articles by Goodall on the Gombe chimpanzees and funded documentary films on her work at Gombe. Nevertheless, while her initial lack of academic credentials was, in Leakey's view, a credential in itself, continued funding came to depend on her willingness to work within an academic framework. While in Cambridge, she often felt that the time away from Gombe compromised her long-term observations of individual chimpanzees, even though, by that time, she had reliable helpers to follow the animals.

Goodall's graduate work and its attendant obligations, however, were the least of her worries about her place in the scientific community. At numerous points in her autobiographical narratives, she comments that, in her early years, she saw her debts to the scientific establishment and the pressure to abide by professional protocols as "ordeals" to be overcome: "Some scientists feel that animals should be labeled by numbers—that to name them is anthropomorphic—but I have always been interested in the *differences* between individuals, and a name is not only more individual than a number but also far easier to remember."[17] Even more ludicrous, to her mind, was the resistance to grammatical gender in scientific writing. When she received proofs of an early technical article, "he" and "she" were struck through and replaced with "it," even though the sex and perhaps even the gender of the individuals were often crucial pieces of information in the description of a behavior. She reinserted the original language and was relieved to discover, when the article appeared in print, that the editor had given up on his attempts to hold that particular line on anthropomorphic discourse.[18]

These were the obstacles, but what of the emerging goals? Always, there was the reality of the chimpanzees themselves and the increasing conviction that Goodall had to present them to the world as they really were—as complicated individuals with rich emotional and mental lives. From Leakey's point of view, Goodall's most exciting discoveries were tool use and meat eating. Goodall was not the first to report these activities, but she was apparently the first to document numerous cases. Several months into her study, she came across David Graybeard sitting on a branch and sharing the meat of a piglet with a female and a juvenile, while an adult bushpig, evidently the parent, paced beneath the tree with her three remaining piglets. Although Goodall did not witness the chimpanzees' cooperative hunting techniques until sometime later, even the omnivorous behavior came as a challenge to the conventional wisdom that these apes were vegetarian. Two weeks later came Goodall's observation of David probing a termite mound with a straw, then popping the insects clinging to the straw into his mouth. She watched David and his companions for eight days during this termiting season before sending a telegram to Leakey, who made the famous remark that the definition of man as a tool-using animal would have to be changed—or chimpanzees would have to be accepted as human. Before long, Goodall discovered that the termites were an important source of protein in the Gombe diet, that infants learned termiting techniques by watching their mothers, and that both tool making and termite fishing required practice. These behaviors convinced her that even if the ability is innate in the chimpanzee, the skill itself must be learned. Thus, termite fishing is also a protocultural phenomenon.

Exciting though these discoveries were, *In the Shadow of Man* focuses on two other aspects of the Gombe chimpanzees' social life: dominance hierarchies—the principal interest of mid-twentieth-century primatologists—and family relationships—a topic investigated mostly by female researchers. Goodall concluded that most chimpanzee males play politics, jockeying constantly for the dominant position in the fairly large but fluid groups that make up chimpanzee society. Dominance carries some advantages in feeding, and dominant individuals are groomed more, but since chimpanzee social organization enables all males to reproduce, a dominant position does not guarantee genetic superiority. Dominance hierarchies are complex: some individual males, such as Goodall's confident and gentle friend David Graybeard, have influence in the community but are content to be second or third in the pecking order, and brothers, notably the sons of the matriarch Flo, often become natural allies who share power. To establish dominance, individuals rely more on violent displays—brandishing logs, lifting the hair, screaming, lunging, and generally running amok—than serious fighting, and injuries from fights are rare. Size obviously gives an ape an advantage in dominance displays, but one especially imaginative individual, Mike, made his point by banging together kerosene cans stolen from the camp. It worked. Yes, chimpanzees are like us.

Furthermore, these early studies suggested that the family bond is essential to chimpanzee life. The emotional connection that forms between mother and infant extends to siblings and grandchildren and is at least as important as the dominance hierarchy. Goodall later acknowledged that studies of other natural groups of chimpanzees had shown that this lifelong bond is not a cultural universal for the species (for instance, most chimpanzee groups rely on female dispersal and thereby avoid inbreeding). But even the integration of this new insight suggests that learning rather than instinct is the source of chimpanzee social structures, just as tool use is a learned behavior. Studying family relationships, most successfully by following Flo and watching her family group at the banana station, enabled Goodall to analyze other aspects of chimpanzee life as well. She made important discoveries about cognitive development, play, grooming, and affective behaviors. Appendix A of *In the Shadow of Man* outlines chimpanzee developmental stages from birth to adulthood; Goodall seems to have been the first to collect enough data to confirm that many chimpanzees are not fully weaned until they are six years old. That being the case, it is perhaps not surprising that, although males and females reach puberty at eight or nine, males do not become socially mature until they are fifteen. Goodall echoes Darwin when she suggests that the chimpanzee's long childhood and late puberty foster family life, and thus the capacity for close affective ties and refined emotional intelligence.

The capacity for cross-species communication is also apparent in the chimpanzee and many other animals. An experience with David Graybeard, related in *In the Shadow of Man* (and elsewhere), reveals this capacity and contributed enormously to Goodall's own emerging goals:

One day, as I sat near him at the bank of a tiny trickle of crystal-clear water, I saw a ripe red palm nut lying on the ground. I picked it up and held it out to him on my open palm. He turned his head away. When I moved my hand closer he looked at it, and then at me. And then he took the fruit, and at the same time held my hand firmly and gently with his own. As I sat motionless he released my hand, looked down at the nut, and dropped it on the ground.

At that moment there was no need of any scientific knowledge to understand his communication of reassurance. The soft pressure of his fingers spoke to me not through intellect but through a more primitive emotional channel: the barrier of untold centuries which has grown up during the separate evolution of man and chimpanzee was, for those few seconds, broken down.

It was a reward beyond my greatest hopes.[19]

This was an iconic moment in Goodall's career.

Cross-species communication for chimpanzees includes communication with other species besides humans. Although Goodall's favorite study species has always been chimpanzees, the olive baboons who congregate around Lake Tanganyika are so numerous that she has been able to make observations of these animals as well. The chimpanzees and baboons have a vexed relationship: although chimpanzees are the larger species, male baboons are about the same size as female chimpanzees and are armed with formidable canines, and the two species often squabble over the same resources. In a conflict, however, male baboons usually give way to both male and female chimpanzees, and infant baboons sometimes fall prey to chimpanzees on a hunt. Some of Goodall's most interesting observations are of adolescent chimps playing with young baboons; adolescents of the two species sometimes engage in experimental copulation with each other, and Goodall describes their fumbling in amusing detail. Quite often interspecies friendships develop between younger animals. Goodall documents one of these friendships in *Through a Window*. A chimp named Gilka and a young baboon named Goblina sought each other out whenever their mothers were in the same area, "gently tickling each other with their fingers or with nibbling, nuzzling movements of the jaws. Their play was accompanied by soft laughing."[20]

Ironically, Goblina's first baby fell prey to chimpanzees, and while Goodall notes that Gilka had no part in these proceedings, she speculates that Gilka might have joined in the feast had she been there. One particular episode involving Gilka, also narrated in *Through a Window*, was a turning point in Goodall's "self-becoming"; it forced her to reconsider everything she thought she knew about her friends the chimpanzees and to rethink some aspects of human nature as well. It is a part of her heroism that she could integrate this shocking new knowledge into a worldview that still allows for hope.

In 1966, a polio epidemic killed or disabled most of the young chimpanzees at Gombe, and the consequences of the plague affected numerous individuals for many years after that, as they tried to cope with motherhood, negotiate the dominance hierarchy, and simply live day to day. Gilka was one of the chimps affected by polio, which left an arm and wrist paralyzed. She had been a beautiful and engaging baby before she was stricken, but the polio was not her only problem: when she was about eleven, soon after the death of her mother, a fungal infection left her face disfigured and interfered with her vision. Goodall speculates that Gilka may have lost a first baby because she was unable to care for it; she certainly did lose one baby, named Gandalf, who disappeared when he was about a month old. In 1974, she lost another, Otta, under bizarre circumstances: Otta was captured and eaten by two other chimpanzees, Pom and her mother, Passion, whose behavior, if it were human, would be diagnosed as sociopathic. Together, these two were observed as they stalked Gilka, snatched at her infant, and finally wrested it from the frantic mother, whose muscular weakness prevented her from defending Otta successfully. After cradling him briefly, Passion killed him in the same way that male chimpanzees kill hunted prey—with a strong, quick bite to the front of the skull. She and Pom ate the infant methodically, the way chimps always consume meat. During this interval, Passion even attempted to caress the grieving mother, who stayed nearby for some time, periodically attempting to retrieve Otta's corpse. Did the empathetic chimp "hardwiring" kick in? Was Passion rubbing salt into the wound? Goodall cannot bear to speculate in print, although most people who have read her book seem unable to resist trying to make sense of this behavior.

Passion and Pom were observed killing or eating five infants, and Goodall estimates that they might have killed and eaten as many as eight altogether. The killings and disappearances stopped only when both mother and daughter became pregnant. As for why they carried out these systematic infanticides, Goodall speculates that the poor parenting skills of both Passion and Pom might have indicated or led to an incapacity for the sympathy and restraint that some of the Gombe chimps seem to have in abundance. Rather than suggesting aggressive survival strategies, Goodall reaches into the tradition of liberal individualism to find an answer to the looming "why." For chimpanzees as well as humans, it seemed to her, the capacity for fellow feeling is a product of nurture more, perhaps, than nature. (In 2006, Christine M. Korsgaard examined such aggressive behavior in response to Frans de Waal's 2004 Tanner Lecture on moral evolution. Korsgaard's conclusion is similar to Goodall's in its liberalism: "The morality of your action is not a function of the content of your intentions. It is a function of the exercise of normative self-government.")[21]

In contrast, it was impossible to dismiss the four-year war at Gombe, also narrated in *Through a Window*, as a failure of individual social skills or a fluke in the chimpanzee social order. The Gombe chimpanzees cohered in three major groups: the Kasekela group, closest to camp and therefore best known to Goodall and her fellow researchers; the Mitumba group to the north; and the powerful Kalande group to the south. The territories of these three groups were distinct, but they were separated by overlap zones, and neighboring groups warily shared resources in these locations. Between 1974 and 1978, war was waged between the Kasekela group and a smaller group that broke off from it to occupy a broad overlap zone between the Kasekela and the Kalande to the south. During this time, Kasekela individuals whom Goodall had learned to love participated in small patrols of four or five males (sometimes with a female named Gigi), stalking, attacking, and killing other individuals in the new group, who had once been close friends and even family members. One by one during this period, the males and most of the females in the new group were attacked and mortally injured, with their assailants hitting, stamping, biting, pulling off strips of skin, twisting limbs, and delivering lethal blows with branches or stone weapons. Such behaviors are never seen between individuals belonging to the same group. It was difficult to determine why the chimpanzees waged this war. But chimpanzees generally do not like strangers, and Goodall finally concluded that the estrangement that arose when the group separated—whatever the reasons for the initial separation—aroused hatred: "By separating themselves, it was as though they forfeited their 'right' to be treated as group members."[22]

Goodall was unprepared for this discovery, but ultimately she integrated it into the larger picture of chimpanzee life that had emerged during her previous fifteen years at Gombe. Up until the end of the 1970s, one of the core themes in primatology was male aggression, as theorized most famously by Solly Zuckerman, who studied male dominance behavior in an unnaturally crowded baboon enclosure at the London Zoo. Zuckerman postulated that human and other primate aggression derives from the same brain structures and from the primate endocrine system and is thus both universal and unavoidable. His research spawned such widely known tomes as filmmaker, playwright, and amateur biologist Robert Ardrey's *African Genesis* (1961), which claims to trace the aggression in our bones through the same fossil evidence that Leakey was uncovering in Africa. Ardrey ventured a little further than the evidence allowed, even at that point in the development of primatology as a science. However, Zuckerman's views on aggression were the orthodoxy for midcentury scientists. They were so powerful that Zuckerman's contemporary, C. R. Carpenter, whose original research on howlers had led to entirely different conclusions, eventually came to ignore his own findings in favor of Zuckerman's.[23]

Goodall did study dominance hierarchies, as we have already seen, but she approached them as part of the dynamics within a large, fluid social system in which friendship and motherhood are also important, and in which dominance has but a minimal influence on mating behaviors. Less may be at stake, it seems, for the male chimpanzee playing king of the mountain than for other primate males, such as silverback gorillas or alpha langurs, who sometimes battle to ensure the perpetuation of their genes and the exclusion of their rivals'. Goodall's work was revolutionary in part because she held this tempered view. But aggression takes other forms, she discovered, and she had difficulty reconciling what she saw during the second decade of her research with the Edenic impression of the chimpanzees she had first created for herself and the public in the 1960s.

The three years between 1974 and 1978 were among the most difficult in Goodall's life. In addition to the students' kidnapping, her grandmother died and she was divorced from Hugo van Lawick. However heartbreaking, these events seem to have been in some ways overshadowed by what she witnessed within the chimpanzee colony itself. Reconciling herself to murder, infanticide, and cannibalism marked a moment of profound "self-becoming." Goodall's journey through that moment was agonized:

During the first ten years of the study I had believed . . . that the Gombe chimpanzees were, for the most part, rather nicer than human beings. I had known aggression could flare up, sometimes for seemingly trivial reasons; chimpanzees are volatile by nature, yet for the most part aggression within the community is more bluster and threat than fierce fighting—a whole lot of "sound and fury signifying nothing." Then suddenly we found that chimpanzees could be brutal—that they, like us, had a dark side to their nature.[24]

Goodall first noticed and called attention to the similarities between the chimpanzees' best selves and our own. After 1978, she began to call attention to the differences. In *Through a Window* and frequently thereafter, she argues a paradox, or, depending on one's point of view, a logical inconsistency: that despite the chimpanzees' cognitive capacity for understanding others' emotions, only humans can be *deliberately* cruel, because only humans are capable of empathy and can fully understand the consequences of our actions. To a degree, this analysis contradicts some of Goodall's other conclusions, and so, to a degree, the problem remains unresolved within her work.

There was some fear in the primatology and conservation communities that lay people would start to wonder why humans should go to great lengths to protect chimpanzees, if the apes are no better than we are. For Jane Goodall, the answer has little to do with the advancement of ecological balance, human knowledge, or even justice, although humans are indeed responsible for the chimpanzees' plight. She could have explained the chimpanzees' violence in terms of evolutionary survival—that real or perceived protein shortages (in the case of Passion and Pom's cannibalism) or food shortages caused by a split between the groups (in the case of patrols and war) account for the chimpanzees' violence toward one another—but she does not emphasize these points. Instead, she finally argues in *Through a Window* that we must protect other species because humans are not the only creatures able to feel happiness and pain, and because our own spiritual survival depends on taking moral responsibility for others: "To fight cruelty, in any shape or form—whether it be toward other human beings or non-human beings—brings us into direct conflict with that unfortunate streak of *inhumanity* that lurks in all of us. If only we could overcome cruelty with compassion we should be well on the way to creating a new and boundless ethic—one that would respect all living beings. We should stand at the threshold of a new era in human evolution—the realization, at last, of our most unique quality: humanity."[25] It is possible that the sense of the paradox can only be rendered clear outside the scientific narrative; in any case, Goodall's nontechnical writings are riven with these and similar concerns.

Although her findings, especially about meat eating and tool making, were exciting for Leakey, Goodall worked toward her own separate goal: healing the breach between humanity and the rest of the natural world. The desideratum quoted above, this goal, and this faith in the ontological value of all beings in nature has remained with her always, even after witnessing the chimpanzees engage in murder, cannibalism, and organized warfare.

IV

But what of coming home? Goodall spent most of her childhood at The Birches, her grandmother's home in Bournemouth, a short walk from the English Channel, surrounded by dogs, poultry coops, and gardens. This home resembled Doctor Dolittle's cottage in Puddleby-on-the-Marsh, or Darwin's lovely Down House, with its generously proportioned grounds, where he wrote his greatest works. But after decades in Africa, Gombe became her home—an endangered home. What she found "out there" made a homecoming like Darwin's all but impossible, and what has urged her forward in the latter part of her career is different from Doctor Dolittle's sense of adventure.

After *Through a Window*, Goodall's science was no longer written in the genre of romance, but her life continues to follow the narrative arc of traditional romance. A kind of homecoming concludes the long, multipart narrative that comprises Goodall's publications, up through the nearly seven-hundred-page 1986 work *The Chimpanzees of Gombe*. This magnificent volume provides an overview of her conclusions about chimpanzee psychology, demographics, communication and relationships, use of territory, social behavior and organization, sex and reproduction, and cognitive abilities. In fact, this encyclopedic book is both the culmination of Goodall's scientific scholarship and an illustrated record of her fieldwork for the general reader. Unlike a novel or a romance, at its core, *The Chimpanzees of Gombe* is a series of biographies—of her best chimpanzee friend David Graybeard, the violent males who put her in her place when she first arrived at Gombe, the successful matriarch Flo, the brilliant leader Mike, the nonparous (sterile) aunty Gigi, the murderous Passion and Pom, and many other individuals. These biographies and the statistical data that accompany them account for much of the scientific value of the book, which covers half a lifetime of a normal free-living chimpanzee who has managed to survive the dangers of childhood. With this book and her previous impressive scholarly publications, Goodall accounts for what was, by 1986, the longest continuous field study of any nonhuman primate, and she sets the standard for subsequent field studies of long-lived land mammals. Perhaps most importantly, Goodall establishes, once and for all, that humans are not the only "subjects" on the planet—that, indeed, we belong to a "communion of subjects," as Paul Waldau and Kimberley Patton describe the community of animal subjects in this world.[26] The emerging idea that drives Goodall's later work is the deepest possible conviction that human good, in this place, this time, and the future, depends upon this spiritual understanding.

The publication of *The Chimpanzees of Gombe* represents the culmination of one phase of Goodall's career and the beginning of another. In the quest romance, the knight errant encounters dangers and temptations; risks death and injury; meets helpers and hinderers; eventually finds knowledge, though usually not in the expected form; and, if lucky, returns sadder and wiser. For most knights errant, the Holy Grail itself is an ever-receding speck in the distance. No questing hero is guaranteed a safe return, but if he does return, his hard-earned wisdom is brought to bear on matters of home, family, community, or state. If Goodall's earlier works together constitute a quest narrative, this big book is not part of that particular project. In some ways, it stands alone, not as a quest but as both an accomplished goal and the turning point in the life and career of one of the twentieth century's most important scientists and activists.

Still, much of her influence derives from the rhetorical power of the quest romance. The Star Trek series replicates the quest memeplex in postindustrial terms: a group of scientists leave home to search for the Holy Grail of knowledge (typically first contact with other sapient species). These scientists errant risk death, injury, and loss, and some of them give their lives in the quest. But the rest find knowledge and experience, and always their ultimate goal is to bring this knowledge back to Earth. In *Voyager*, the most poignant of the Star Trek series, the starship crew become trapped in an alien galaxy when the singularity they have ridden across the universe disappears. Until she can find a way back to Earth, Captain Kathryn Janeway—the echo of Jane Goodall's name is perhaps not accidental—must bear her own grief as well as that of her crew. For herself and them, she must learn how to be at home in this alien place, while still attempting to return to Earth and all she has lost. She is adrift in time and space, and she has a mission.

Janeway's dilemma is not unlike Jane Goodall's. The risks, dangers, ecstasies, and discoveries of Goodall's voyage out are chronicled in scientific papers, *National Geographic* articles, acclaimed books written for both the general public and the scientific community, and her 1986 volume. Her work afterward is chronicled less in scientific narratives—although she has continued to write and contribute to scientific publications—than in spiritual autobiographies, impassioned pleas for environmental responsibility, and more than a dozen books for children, which, like the ones that inspired her own childhood, tell the stories of animals.

The vast tropical forests that Goodall could see from the peaks of Gombe in the early years of her stay there are cleared, cultivated, or spoiled, right up to the edge of the park. The political landscapes of central Africa put all animal species, humans and nonhumans alike, at risk—and, indeed, the pressures on habitat for all species are global and urgent. Goodall explains in *Harvest for Hope* (2005) that at a scientific conference in 1986, in the midst of celebrating her greatest scientific achievement, she found herself engaged in a dialogue about the critical need for conservation as well as research. Her career and her life took a sharp turn. She became an advocate for animals confined in biomedical or behavioral laboratories, zoos, and animal parks; abused as pets or entertainers; illegally killed in the bushmeat trade; or endangered by habitat loss and human overpopulation.

In *Harvest for Hope*, Goodall turns her attention to the critical problem of feeding the world responsibly, humanely, and sustainably. As she sees it, the crush of human misery, brought about by the "unsustainable lifestyle of the elite societies around the world," is both the cause of these terrific pressures on the wildlife she loves and an ethical mandate in itself.[27] For Goodall, the fate of our own species is bound up with the fate of the planet which is our home and that of all the life-forms sharing it. Like Captain Janeway, who can no longer focus on research and yet cannot return to Earth, Goodall cannot stay in Africa with the chimpanzees or finally return to the known comforts of the industrialized world; instead, she is adrift in time and place, and she has an urgent mission.

Her example has been an inspiration to those coming after her, as well as a challenge.

CHAPTER THREE

Tragedy of the Field

Looking down from above, I saw the
fragility of that small area I called *the field*, saw
once again how interwoven it was with
human needs and problems.

—SHIRLEY STRUM, *Almost Human:
A Journey into the World of Baboons*

I

The "field" is a real place, but it is also, paradoxically, an
expression of the human imagination. It is an area selected
within the natural range of a particular nonhuman species or,
in the case of cultural anthropology, a particular (stereotypically preindustrial) human society. But the field is bounded
artificially from the outside; individuals and collective entities
within it are named and classified from an outside perspective. The field exists in linear human time, not earth time; it is
mapped from a human perspective; and it is remade from the
inside, the space invariably altered by technological presence,
the camera flash of the human eye. The field is a special construction—a chronotope, or time-space, as Mikhail Bakhtin
says of setting in the novel. It is perilous for scientists who

work and live in the field to forget that they are surrounded by their own artifice.

This is true whether the human imagination that conceives the field is scientific or literary. Here is the field established by a famous primatologist, described by a famous poet:

> Completely protected on all sides
> by volcanoes
> a woman . . . sleeps in central Africa.
> In her dreams, her notebooks . . .
> the mountain gorillas move through their life term
>   [and] . . .
> inhabit, with her, the wooded highland.

Imprisoned within the daily grind of industrialized patriarchal society, the speaker expresses envy of "the pale gorilla-scented dawn," "the stream where she washes her hair," and "the camera-flash of her quiet / eye." In 1968, on the basis of a newspaper sketch, Adrienne Rich thus described Dian Fossey's life as a romantic idyll and the field as Arcadia in her poem "The Observer." But as the cliché has it, "et in Arcadia ego": death, too, is part of the green world.

It is difficult now to know what to think or how to feel about Dian Fossey. The idyllic view of her early years in the field—created by the images that inspired Rich's poem, expected in the wake of Jane Goodall's experience, fervently wished for by Fossey herself, and even suggested by one of the names for her geographical location, the Mountains of the Moon—conflicts with a reality all too situated in history. At the time that Rich constructed this idyll, the ironies of Fossey's life could not have been predicted from the information available to the ordinary American newspaper reader. It did not convey the violence that marked her personal interactions with others, the threats to the mountain gorillas' existence by poaching and habitat loss, the false calm of a forest that would soon be at the center of waves of genocide, or even the violent potential of the gorillas themselves. Only later, after Fossey's articles started to appear in 1970 in *National Geographic*, would the ironies of her life and career gradually become apparent. Before that, it would have been difficult for anyone to detect danger in her Arcadia.

If the stories Fossey wrote in *Gorillas in the Mist* and for *National Geographic* reveal a scientific methodology and world-view similar to Jane Goodall's, Fossey's death ripped the narrative out of her control, and in media, popular culture, and biography, it became not the idyll that Rich imagined but a tragedy. In classical and Renaissance tragedy, disaster is brought about partly through the hero's fatal flaw. Fossey's fatal flaw was an inability to compartmentalize as a scientist must. Although in primate study it is difficult and perhaps not even desirable to separate science and emotion, it is important to *know* the difference, and to know the extent to which desire can interfere with objectivity. Had Fossey been able to look at her situation with greater objectivity, she might not have blurred the boundary between "home" and "the field" quite so much.

Fossey was, in fact, never safe in the Mountains of the Moon. The gorillas were threatened by poaching and habitat loss, and according to her critics, who are strident and many, Fossey's attempts to protect them may have endangered the gorillas further by alienating local residents and government officials, who then refused to cooperate fully with her conservation efforts. Her approach to conservation may have endangered even her own life. Finally, although she was the strong, independent female Rich assumed her to be, there was nothing quiet about the flash of Dian Fossey's eye. Ever since her early years at Karisoke, Fossey has been perceived by some as a villain and by others as a hero, but almost no one occupies neutral ground in the controversies surrounding her work.

It may seem surprising that a poet as politically astute as Rich could not avoid the conflation of wilderness with purity, and the configuration of the scientific field as a neutral or innocent space. But why should Rich have understood what Jane Goodall was just then learning through her own harrowing experiences, and what Dian Fossey was to learn only after years of struggle—that the field is a humanly constructed, morally implicated, perilous space from which the hero returns, or risks exile?

Fossey's view of her field was controversial, but it was also distinctly Western, and that was part of the problem. In the beginning, Western field scientists failed to take into consideration the degree to which they themselves constructed the field, and why. Japan and India have resident monkey populations and a tradition of scientific primatology, and in recent decades, there has been a more pronounced Asian presence in the discipline; scientists from other parts of the world where monkeys live now study them as well. But for the most part, this has largely been a Western discipline with imperialist roots. Donna Haraway's many essays on primatology, collected mostly in *Simians, Cyborgs, and Women* (1991) and *Primate Visions* (1989), constitute a fine history of American primatology, in particular, including its beginnings in the work of American psychologist and physiologist Robert Yerkes. Yerkes's research grew partly out of his involvement in the eugenics movement, and he developed animal experimentation in the laboratory to inform his social engineering projects, which included intelligence testing for the armed forces during World War I. These tests were later used for race research and immigration restriction (a reminder that, like Poe, early observers of nonhuman primates seemed to inevitably link their thoughts about these animals with race in human populations). In carrying out his experiments, Yerkes worked with many species of captive primates, but never in the field.[1]

Although laboratory research and fieldwork complement each other, fieldwork is quite unlike laboratory science, not least in its effects on the researcher. That distinction will become more striking in chapter 6, about the work of the baboon researcher Robert Sapolsky, whose fieldwork was begun as a complement to experimental research in the laboratory. Without quite articulating the constructedness of the "field," scientists such as Fossey saw themselves and their methods mainly in opposition to the laboratory and, of course, in opposition to local pressures on animal habitat. In Darwin's day and well into the twentieth century, primatology fieldwork consisted mostly of collecting, an arduous and time-consuming pursuit. Now, as then, Europeans and North Americans who work with primates outside a laboratory or other facility for captive animals must travel, because there are no free indigenous nonhuman primates in Europe or North America (if southern Mexico is excepted).

The longer the fieldwork lasts, the more the scientist risks exile—and that is certainly what Fossey experienced. As Michael Ugarte puts it, "Exile is precisely that shaky middle or liminal ground which tempts us to forsake the comforts and routine of the home in search of an elsewhere. Exile . . . begins with indignation and ends (if it ends) with questions."[2] In Fossey's case, the reverse was true. Her exile was not only spatial but oddly temporal, since great apes (and many other primate species) are adapted to rough and remote terrain presumably because these were the only domains left to them by the dominant hominids who first deserted the forests for the plains more than a million and a half years ago. Field researchers also face other psychological pressures unknown to early collectors and adventurers. For instance, since most primatologists are drawn to their subjects by profound sympathy or profound curiosity about an "other" who is such close kin, they cannot and probably should not avoid emotional involvement with their study animals. For Fossey and "her" gorillas, it was "us against them."

Field scientists also inevitably pack the imperialist baggage of their discipline, which can hinder their work, and they have to contend with the results of imperialism: postcolonial political confusion left in the wake of European withdrawal from the global South, habitat loss suffered by their study species, economic disarray brought about by the politics of extractive practices and economic development, and decimation of animal populations, in part by the "great white hunters" and collectors whom the scientists have followed into the field and more recently by the destruction of habitat and the bushmeat trade. Women in primatology have had to endure even more struggles than men, since the culture of Western science has traditionally stacked the deck against them, and since they are at greater risk for disrespect, or worse, at home and abroad.

To be sure, as human populations expand and development continues, field sites are more connected and comfortable than when George Schaller first explored the Virungas. Nevertheless, Fossey found herself in a curious, liminal position: she was a voluntary exile in a field that seemed strangely familiar and yet was sometimes dangerous. The animals she studied hold a mirror up to human nature, and their very existence is endangered. So, too, is the existence of the field. The field may be defined by scientific protocols, but it is constructed of myth, dreams, popular culture, and practical human needs, as well as data.

No wonder coming back is a greater challenge than the voyage out. Dian Fossey's extreme passions and unsolved murder ultimately cast her as a tragic exception within the profession, but her attachment to the animals and the difficulties of her life and work are exemplary.

II

The field is established within an animal's range. Within the historical record, three species of gorilla—the western lowland gorilla (*Gorilla gorilla*), the eastern lowland gorilla (*Gorilla graueri*), and the mountain gorilla (*Gorilla beringei*)—have lived in small pockets across equatorial Africa. These pockets have been relatively isolated from one another and inaccessible to humans, and since gorilla populations have been small in comparison to other primate populations, it is not surprising that the mountain gorilla was not "discovered" until 1903, by the German zoologist Paul Matschie (a tree kangaroo was named after him, but not the mountain gorilla, who is in fact named for a hunter). Nor is it surprising that even in the nineteenth century, some European and American observers worried about pressures from hunting and habitat loss and feared eventual extinction for gorillas.

Gorillas live in small groups, or families, a social structure that suits them because not all of their range provides adequate habitat—specific locations of the resources an animal needs for survival—and thus the gorillas travel together from place to place, a practice that also seems to protect them from external parasites. In evolutionary terms, humans are the most successful primates because we are generalists; our range is the whole earth, and human habitat can be found almost anywhere, even in a space station floating above the earth. Other primates rarely alter their whole range to consist of habitat. Mountain gorillas are among the most specialized of the primates, and each group needs a relatively large range that offers plenty of foods in the mountain gorilla diet: occasional grubs and snails, bamboo shoots, thistles, nettles, wild celery, pygeum fruit, mistletoe, blackberries, fungi, and large quantities of galium vine. Some of these foods are seasonal. Mountain gorillas move around almost every day, partly in search of food, and they build new sleeping nests each night. Thus, they require woodland with tall trees for lookouts and nesting materials. Since most of the gorillas in a group will sleep at ground level, they also need enough understory vegetation for sleeping nests and the privacy they seem to crave.

The range size of the gorilla population is linked not just to the availability of foods and other resources but also to the social need for each group to have a well-defined territory. Although these territories may overlap in times of seasonal scarcity or when part of a territory is destroyed, boundaries are diligently enforced. In the Virungas, where about half of the free mountain gorillas now live (the rest live in Uganda's Bwindi Impenetrable Forest, which has recently become a national park), gorillas sometimes travel through the highest parts of the mountains, but these areas are too cold for them to stay long, and foods are scarcer at higher altitudes. Even though the higher elevations are part of their range, these sections of the protected zone do not provide gorilla habitat. So, when part of their range is lost, the mountain gorilla population may sustain a stress level out of proportion to the size of the lost area. Normally, only small, scattered areas within the gorillas' range actually provide suitable habitat.

The western lowland gorilla, the most numerous species, now ranges through several countries in west central Africa, while the eastern lowland gorilla lives in part of the Democratic Republic of Congo. The mountain gorilla's range has always been small compared to the ranges of its lowland cousins, but in the middle of the twentieth century, it included 35,000 square miles to the west of the Rift Valley. These gorillas inhabit a mountainous region that includes the Impenetrable Forest and the Virungas, a corridor of mostly inactive volcanoes. In 1925, a protected area of mountainous terrain was established in Rwanda. Between 1925 and 1929, the Belgian government established the Albert Park in the same region, in what is now the Democratic Republic of Congo. And in 1939, contiguous with these protected areas, a mountain gorilla reserve was established near Kisoro, Uganda. This multinational gorilla zone is now known as the Parc des Volcans in Rwanda, the Parc des Virunga in the Democratic Republic of Congo, and the Kigezi Gorilla Sanctuary in

Uganda. By 1950, the forest was gone, right to the edge of this zone, and since then the boundaries of the remaining gorilla areas have been scaled back more than once to meet the demands of exploding human populations, especially in Rwanda.[3]

The mountain gorilla has been studied in all three countries, but nowhere as intensively as at Karisoke. This research station was established by Dian Fossey in Rwanda, between Mount Visoke and Mount Karisimbi, inactive volcanoes that straddle the Rwanda-Congo border. However, Fossey began her work about four miles away, on the other side of the border, at the location selected in 1959 by Schaller, who conducted the first long-term field study of the mountain gorilla. Ironically, Schaller's cabin was built decades earlier by Carl Akeley, a hunter who supplied specimens for the African Hall of the American Museum of Natural History in 1921, fell in love with the animals he killed, helped persuade the Belgian government to set aside Albert Park for their protection, and died when, despite failing health, he returned to Africa. He was buried outside the mountain cabin.

When Schaller and his University of Wisconsin professor John Emlen surveyed the mountain gorilla population, they estimated a total of 8,000 to 9,000 individuals (with a possible low of 5,000 and a possible high of 15,000), and they discovered that the common name "mountain gorilla" was actually a misnomer. At that time, three quarters of the population favored forests at elevations of less than five thousand feet. Twenty years later, when most lower-altitude mountain gorilla habitat had been destroyed, the name was more fitting. By the time Fossey published *Gorillas in the Mist* in 1983, the safest part of the mountain gorilla's range had become artificially confined to the protected areas of the Virungas, and she estimated the population in those areas to be 240—a drastic and dangerous reduction. During her stay there, at least one group of gorillas responded to territorial pressures by moving so far up into the mountains that they were threatened with respiratory illnesses and inadequate diet. In *In the Kingdom of Gorillas*, Amy Vedder and Bill Weber report that by the 1990s, even though a decline in poaching resulted in a slight population increase within the Virungas, the gorillas' range had shrunk further, making the existing population vulnerable to genetic isolation and inbreeding.

After the Rwandan genocide and disturbances in Uganda in the early 1990s, Schaller reported in a 1995 article for *National Geographic* that the total mountain gorilla population hovered around 600 (about 320 in the Impenetrable Forest and 300 in the Virungas). But the pressures in this area have not subsided.[4] In a more recent *National Geographic* article, Mark Jenkins reported that seven gorillas were murdered for obscure reasons in July 2007 in the Congolese section of the park, where the latest threat to habitat is the burgeoning charcoal industry, in spite of dedicated attempts by park personnel to protect the area. Between the date Schaller's article was published and the period of Jenkins's reporting in 2008, more than a hundred park rangers were killed.[5]

The field is an area within the animal's range. Sadly, in recent years, the field for the study of mountain gorillas has overlapped almost precisely with their full range. This was not originally the case. Schaller's original study site in the Virungas was chosen in late 1959 after Walter Baumgärtel, an adventuresome gorilla enthusiast who ran a small hotel in Uganda called Travellers Rest, petitioned Louis Leakey to find someone to study the local mountain gorilla population. Of course, this project coincided perfectly with Leakey's own agenda. But Schaller and Emlen did not settle on the Virungas until completing their census, which revealed that the population density of the mountain gorillas was highest there.

Like many field primatologists, Schaller felt constrained by the form of scientific publication because he had so much more to say. In *The Year of the Gorilla*, he writes, "I had no space to reveal the enjoyment I derived from roaming across grassy plains and uninhabited forests and climbing mist-shrouded mountains."[6] The cabin occupied by Schaller and his wife, Kay, was situated in a meadow surrounded by mountain peaks, and it was breathtakingly beautiful:

Occasionally, when I stood in the sodden forest, muddy from the waist down and with all my clothes wet through, it fleetingly occurred to me that somewhere there must surely be an easier and climatically more inviting place to study. But then I looked about me and saw the massive bases of the mountains, and the forest of gnarled trees with a light like muted silver filtering through the tresses of grey-green lichen that festooned the branches. This Elysium was mine, shared only with Kay and the gorillas and the other creatures; "climb the mountains and get their glad tidings, and nature's peace will flow into you . . ." wrote John Muir, and I knew what he meant. I would not have traded Kabara for any place on earth.[7]

Aesthetic awe, gratitude, and ownership: these are the emotions expressed in Schaller's comments. He knows that this gift is temporary, but he would not trade his situation to be anywhere else.

Researchers in the field do not attempt to study every individual in the study area; instead, they choose focal individuals or groups to follow and study. In gorilla studies, researchers concentrate on the family group, which averages about a dozen members and is the basic unit in gorilla social life. Typically, in the Virungas, researchers have based most of their research on those easiest to study. Sometimes this means the groups that are easiest to habituate, but most of the study groups are simply those whose home ranges are most accessible to the researchers and nearest their base camps. Researchers keep records of everything they know about their study groups, from parasites found in dung to play behaviors. By far, the most exhausting part of this research is mapping the gorillas' routes through their habitat, including stops to eat and rest, sleeping sites, locations of encounters with other groups, and locations of other important events in the gorillas' lives.

In 1959–60, Schaller managed to follow one group for a period of over six weeks, mapping their night stops, and *The Year of the Gorilla* includes a map of this labyrinthine route on the slopes of Mount Mikeno. Following the group over and around treacherous cliffs and gullies, through dense forest in rain and fog, required great effort. (Decades later, when Robert Sapolsky took a break from his baboon research to follow in the steps of Schaller and Fossey, he described footing it through this terrain in more frightening, and hilarious, detail.) Mapping the nesting sites of individuals when they stop for the night, as Schaller did, is difficult because the nests can be as much as fifty feet apart, and they are meant to be concealed.

Such mapping is partial: it accounts for what the gorillas do when they can be found, on a day when the researcher has time to look for them, in places where the researcher can follow. Enough repetitions, enough maps, and enough records theoretically result in an accurate picture of the animals' lives. But careful scientists and dedicated animal observers know that the picture is always incomplete, and their conclusions are reached inductively instead of with the certainty of experimental science. Consequently, the typical response of the scientist and researcher, especially one in love with the subject, is a more or less obsessive desire to track the gorillas every hour of every day. The researcher goes native, temporarily or permanently, as Schaller describes:

Hours passed. The rain continued to pelt down on my back, and the gorillas remained in their shelter watching me mutely. There is a tale that on Christmas night man and animals forget their differences and can converse with each other on equal terms. The gorillas talked to me at times with their expressive eyes, and I felt that we understood each other, but no sounds passed our lips. The forest darkened imperceptibly, and with a shock I realized that it was five o'clock. I left the gorillas and hurried home. When I emerged from the forest at the far end of the meadow, I saw Kay standing on the rise by the cabin, waiting for me, looking very lonely in the fog and gathering dusk. Tears were streaming down her face, and when I held her close, she told me that it was very late, it was Christmas, and she thought I had been injured because I had not come home.[8]

Schaller goes on to list what he and Kay had for their Christmas dinner; he expresses no more remorse for frightening her than the little implied in this passage.

According to Schaller's introduction, his pilgrimage to Africa came about from "chance remarks and luck." His account suggests that he was happy with his adventuresome wife and his graduate studies in bird behavior, and his career took a turn toward gorillas only when Emlen heard about Baumgärtel's petition second- or thirdhand. When Schaller left home for the Virungas, however, his imagination was fired by the tales of unstudied gorillas, which had inspired generations of explorers before him, and he fell in love with his subjects.

Schaller had to end his study after a year, not only in keeping with the terms of his appointment but also due to political unrest in the region. When he returned to conduct a short study the following year, he did so with some trepidation and without Kay. But there is no doubt that he felt tempted by the wilderness and by the gorillas, and that he returned home with more questions than answers. Even after a visit in the 1990s, the questions persisted: Do gorillas understand their evolutionary kinship with us? How can their species survive in a land of crushing poverty and ethnic violence? How have they survived this long?

III

Dian Fossey, Schaller's successor in the Virungas, readily acknowledged her debt to him, but in the almost two decades of her stay there, the research done at Karisoke surpassed, by far, what Schaller's study had yielded. Fossey was a different kind of exile. Driven by a profound interest in animals and Africa, she was first fired up to visit the gorillas at an address Leakey delivered in Louisville, Kentucky, where she worked as an occupational therapist. Fossey recognized the value in Leakey's project of filling gaps in human evolutionary history by working backward from apes, but she had many other reasons for going to Africa. She came to the project not from a specialized graduate program, but from working with children in an occupational therapy program. And although her background in human physiology was ultimately helpful in her gorilla research, she studied Swahili and the primatology literature on her own before she felt qualified to undertake the research.

Fossey's journey to Africa was fueled not just by curiosity about the gorillas. Africa may well have provided, in addition, the prospect of escape. Farley Mowat's biography *Woman in the Mists* suggests that Fossey's move across the country from California to Kentucky could have been the first phase of a journey away from an unsympathetic and dysfunctional family. Even though the African experience was immeasurably more than an escape from home, her background may have given her the strength to endure the loneliness. As a field biologist studying the arctic wolf, Mowat was eminently equipped to write about Fossey's life; removed from the passions and politics of the scientific establishment, he still understood from personal experience what it was like to work in extreme conditions, endure extreme isolation, and mourn the deaths of animals.

Like Jane Goodall, Fossey was influenced by children's stories. Significantly, though, her model seems to have been not the doughty Doctor Dolittle but the film based on Marjorie Kinnan Rawlings's novel *The Yearling*, in which the main character, Jody, the child of an impoverished family in Florida's Ocala Forest, adopts a wild orphaned faun. Among Fossey's most profound emotional experiences in the Virungas were caring for orphaned animals and then having to let them go. Her childhood preference for stories is reflected in both her early career choice and the force of her impulse to protect and preserve the forest animals later on.

Living on a farm outside Louisville, surrounded by pets and working with disabled children, Fossey was living her life in a perfectly functional, almost ordinary way when she met Louis Leakey. She wanted more than emotional survival, though. Like Schaller, Goodall, and many other field biologists, Fossey seems compelled in her field narrative *Gorillas in the Mist* and elsewhere to answer an almost metaphysical question: how she came to be in the field. Mowat's biography, composed of his own narrative interwoven with long passages from Fossey's unpublished diaries and letters, begins with her words:

Neither destiny nor fate took me to Africa. Nor was it romance. I had a deep wish to see and live with wild animals in a world that hadn't yet been completely changed by humans. I guess I really wanted to go backward in time. From my childhood I believed that was what going to Africa would be, but by 1963, when I was first able to make a trip there, it was not that way anymore. There were only a few places other than the deserts and swamps that hadn't been overrun by people. Almost at the end of my trip I found the place I had been looking for.[9]

At first glance, this story echoes Jane Goodall's, but more than Goodall's, these are the words of an exile—not an exile from the vagaries of misused governmental or military power, but from the politics of the family, from the power plays within its affective ties, and from time itself. Fossey did not know it at first, but the Virungas, for all their seemingly pristine beauty, were traditionally inhabited by the Batwa and marked by the Europeans who had come before her. Within decades, one of the busiest routes for refugees would wind through these mountains, and fields at the foot of the mountains on the Congolese side, within a few hours' walk of gorilla habitat, would be packed rock-hard by the feet of hundreds of thousands of refugees encamped there or traveling through.

Without institutional support, Fossey borrowed money for her first journey, a seven-week safari that included a visit to the Leakeys' research site in Olduvai Gorge and gorilla watching. After that, she wrote, "there was no way that I could explain to dogs, friends, or parents my compelling need to return to Africa to launch a long-term study of the gorillas."[10] Soon, her purpose for conducting scientific research and earning her Ph.D. with Goodall's professor Robert Hinde also became clear: science was necessary to save the gorillas. "Why do gorillas go where they do?" asks Fossey in her second article for *National Geographic* in 1971. "Do their routes remain stable or do they vary? How many still survive? What is their present territory? I have compelling reason for wanting to know. If we are to save the animals from extinction, we must find answers to these vital questions. We must learn the areas of known population concentration before we can provide protection—and thus my interest in both an accurate census of gorilla numbers and a study of gorilla ranges."[11]

Like Goodall, Fossey directed her first book (as it turned out, her only book) at both a scientific and lay audience. It is replete with valuable scientific information, with most of the hard data and technical descriptions relegated to a series of appendixes. If the professional scientist among Fossey's readers has no need for a rich description of setting, that is not true for the lay reader, for whom she wrote the book as an autobiographical travelogue. Fossey is an engaging writer who describes the mountain gorilla's homeland in loving detail.

After six months in Uganda, Fossey saw endemic political violence escalate. She was forced out of the location chosen by Schaller, imprisoned, and probably raped before escaping, although the story she tells about this episode in *Gorillas in the Mist* is understandably stripped of details, since the narrative emphasis is on the field, the gorillas, and her own engagement with them. She quickly set up a second camp, just across the border in Rwanda, in another area of magical beauty. Following a difficult journey up from the base of the mountains, she writes,

we continued climbing at an easier gradient for more than an hour before coming to the beginning of a long meadow corridor densely carpeted with a variety of grasses, clovers, and wildflowers. Distributed throughout the meadows, like so many powerful sentries, stood magnificent Hagenia *trees, bearded by long lacy strands of lichen flowing from their orchid-laden limbs. The entire scene was backlit by sunlight, giving all a specular dimension no camera could record or eye believe. I have yet to see a more impressive spot in all of the Virungas or a more ideal location for gorilla research.*[12]

The area described here is about four miles east of Schaller's original location. Fossey called it Karisoke, combining the names of Mount Karisimbi and Mount Visoke, and the gorilla groups nearest Karisoke became her most important study subjects. *Gorillas in the Mist* is the story of Fossey's life in this setting, as well as the lives of fifty-one individuals, in Groups 4, 5, 8, and 9, whose range consisted of a nine-square-mile area around the camp. Fossey, the local men employed as trackers and camp staff, and, over the next eighteen years, a number of students who came to conduct their own research and contribute to the long-term study at Karisoke thus occupied a field that was originally chosen by Schaller on the basis of gorilla population density. Karisoke, specifically, was chosen as a campsite in part because it was a beautiful and practical place to pitch tents and eventually build cabins. By the end of Fossey's story, there were seven cabins, two fire pits for burning fallen branches, toilets, a kitchen shed, a henhouse, a weather station, a cemetery, and numerous gravel paths. Today, paths also link the camp with other areas in the Virungas, including a car park that empties onto the road leading to park headquarters and, a bit farther, to Ruhengeri, the nearest Rwandan town. No field research has ever been conducted in a place more situated in history, politics, local and world economies, stories, myths, and suppositions. Yet the difficulty of reaching the research station and the emotional pressures of working with such a congenial and threatened species soon forced its human inhabitants into a vexing combination of comradeship and strife, which became more intense with the passing years.

Fossey thought of the forest as her "real home" and the men who staffed the research center as her family.[13] And it must have seemed natural, once she had habituated the normally gentle and peaceable gorilla family groups—mothers, children, infants, adolescents, a silverback, and perhaps one or more subadults—to think of them, too, as family. Quite early in her research, Fossey realized that accurate observations could be made only if she minimized her intrusions on the gorillas' way of life. Her method of habituating owes much to Schaller's accounts and to Goodall's advice about imitating ape behavior to gain their trust. In fact, Fossey sometimes made an entrance at her lectures by knuckle-walking onto the stage, and Goodall still begins speeches with a series of spine-tingling pant-hoots. Both scientists have been criticized for choosing to conduct such close observations, rather than watching at a distance (and learning much less). But both were well aware of the dilemma, realizing that close observation might, in fact, change the very behavior they wanted to document. Because of the gorilla's slower, gentler, and more intimate social behavior, Fossey was able to decrease the distance between herself and her study animals even more, or more consistently, than Goodall did, and without a feeding station.

As method actors know, consciously imitating behavior results in an intuition of the behavior that works from the inside out, and thus a better performance. Imitating apes in order to observe them closely leads to an intuition of apeness. Dawn Prince-Hughes, who has Asperger's syndrome, considers the captive lowland gorillas with whom she works to be language-less persons, closer to her in many ways than *Homo sapiens*. For Prince-Hughes, habituation means relying entirely on her own intuitions as she imitates ape behavior.[14] Since most of us unconsciously reflect the behaviors of those around us—accents, tics, gestures—perhaps becoming animal is inevitable for someone who spends as much time around gorillas as humans. Habituation works both ways. Sometimes the researcher goes native and the field becomes home.

Fossey's accounts of lazy gorilla days are filled with humorous descriptions of gorilla games, facial expressions, and body language. She casts herself as an odd woman out. On her first visit to the gorillas, the guide had to explain that she should be tracking in the direction the gorillas were moving, not trying to figure out where they had been! In fact, the joke was always on her: she describes herself pretending that wild celery was her favorite food, being whacked with a bundle of leafy branches by a playful subadult trying out dominance behaviors, watching helplessly as an adolescent ate her field notes or an infant absconded with a camera, tripping on vegetation, being ousted along with another female from a hollow tree coveted by a silverback during a rainstorm, and once melting with terror at a silverback's charge—which mercifully stopped less than three feet away. Every day with the gorillas was rewarding, but perhaps her best day came early in her research at Karisoke:

The first occasion when I felt I might have crossed an intangible barrier between human and ape occurred about ten months after beginning the research at Karisoke. Peanuts, Group 8's youngest male, was feeding about fifteen feet away when he suddenly stopped and turned to stare directly at me. The expression in his eyes was unfathomable. Spellbound, I returned his gaze—a gaze that seemed to combine elements of inquiry and of acceptance. Peanuts ended this unforgettable moment by sighing deeply, and slowly resumed feeding. Jubilant, I returned to camp and cabled Dr. Leakey I'VE FINALLY BEEN ACCEPTED BY A GORILLA.[15]

Although Fossey's pioneering work has been foundational in the understanding of gorilla behavior and ecology, it is probably significant that, for her personally, the breakthrough moment was not the discovery of a protocultural behavior, but the beginning of an emotional connection that would become richer and more powerful as the years passed. Forever after, Fossey regarded Peanuts as a special friend, though her love for him was overshadowed—not least in media representations of Fossey's project—by her affection for Digit, a young silverback killed by poachers.

Of the many criticisms of Dian Fossey, comments about her interactions with poachers have been the loudest, not only because she sometimes seemed indifferent to the needs of the local people but also because, in the view of some wildlife preservationists, her treatment of poachers invited retaliation against the gorillas themselves. Hunting has traditionally been important to the diet of the Batwa, who have inhabited the Virungas and the surrounding area since long before Europeans came to explore and exploit the countryside. In *Gorillas in the Mist*, Fossey acknowledges that the Batwa themselves—indeed, all of the Africans living near the parkland—are subject to political and military pressures from other ethnic groups and population pressures that make traditional ways of life hard or impossible to sustain. The Batwa have not deliberately killed gorillas, but sometimes gorillas are accidentally caught in traps set for small antelopes, called duikers, or other game. Fossey was livid whenever a trap entangled a gorilla; even if the animal escaped, the trap wire could become embedded in flesh or result in a severe injury or fatal infection. For almost a half century, since observations of mountain gorillas began, researchers have repeatedly noted individuals missing fingers, hands, and feet. Fossey attacked poachers in any way she could: she destroyed traps and frightened them with threats of retaliation, gunshots into the air, and *sumu* (indigenous magic practices). On occasion, she detained them. She reasoned that gorillas cannot protect themselves individually or as a species on the brink of extinction. However, most of her critics argue that Fossey's methods were ineffective in the long run because they alienated local people, and some of her critics interpret Fossey's vengeance as racist.

The most likely groups to retaliate, however, were not the game trappers but those who set out to capture infants for the live animal trade. Every capture meant the killing of adults trying to defend the infant, and captive infants coped badly with confinement and the loss of their mothers. The captors were frequently ignorant or indifferent to their basic needs, so these infants usually died. The famous Coco and Pucker, whom Fossey rescued from dealers, nursed back to health, and reluctantly relinquished to the Cologne Zoo, provide merely one example of how difficult it is to maintain mountain gorillas in captivity, even when the infant survives a long and terrifying journey. Both gorillas died in 1978, after less than nine years away from the mountains. This kind of deliberate poaching led to one more indignity: dead gorillas' heads, hands, and feet were often taken as trophies, for which there was, at that time, a ready tourist market. Sometimes their bodies were consumed as bushmeat. Since the international CITES convention went into force in 1975, accredited zoos and research facilities no longer purchase endangered animals unless they are born in captivity, but private collectors have had no qualms about disobeying these proscriptions, and the bushmeat trade has actually grown as more logging and mining roads have appeared and local taboos have faded.

Fossey met Digit as an infant soon after she arrived in the Virungas, and by the time he died on New Year's Eve 1977, he was the second-ranking male in his group. Digit had just reached breeding age and impregnated one of the females in the group. He was tolerated by Whinny, the dominant silverback, because he supported the patriarch and took on the role of sentry. Fossey noted that in the months before his death, Digit seemed somewhat isolated by this role, and she felt a kinship with him because he was turning into a loner. Had he lived, Digit would have assumed leadership of his group if Whinny had soon died or become disabled, or he would have separated from the group to form a new family, or he might have socialized with a bachelor band for a period of time. But as long as a gorilla remains with a family group, he will aid and protect its members and fight to the death to defend them. Digit was killed by at least five spear thrusts while his family group escaped from poachers. Since no infants were missing in the count made after his body was found, the motives of the poachers are a matter of speculation. Fossey's best guess was that Batwa poachers met the gorillas while inspecting their traps and reacted to a charge by killing Digit, returning later to cut off his head and hands with a panga, or

African machete. No doubt buyers for these "trophies" were found.

Because Digit's image was already recognized from travel posters, his demise was international headline news. He was buried next to the cabin, where he continued to remind Fossey of the peril faced by his kind and of family loyalty—to the death. Digit's death was a turning point in her life. She already hated the poachers, loved the gorillas as family, and shored up the fragments of her life at Karisoke against the threat of ruin, but she became even more absorbed in saving the apes' lives afterward. Her relations grew more strained with almost everyone else, from the local people and Rwandan government officials to wildlife protection agencies in Europe and North America, from the students conducting research at Karisoke to university professors and the National Geographic Society. Board members for the Digit Fund, which Fossey established to fund antipoaching controls, felt that the money raised by publicizing Digit's death would be better spent on developing gorilla tourism—almost as harmful to the animals as the poachers, Fossey believed, because it altered gorilla behavior and exposed them to human diseases. As funding sources dried up or were tied up in the hands of others, Karisoke became more difficult to manage; when Fossey had to leave, as she did from time to time in response to the demands of her work, it became harder to negotiate the management of the center in her absence. It also became harder to obtain a visa to remain. The more endangered the gorillas were and the more vulnerable Fossey felt, the harder it was to take effective action to protect herself. Her life grew increasingly chaotic. Ties to the United States stretched thin, her position in the mountains with the gorillas became tenuous, and external and internal defenses almost disappeared.

Fossey's trajectory can be charted by comparing the three articles she wrote for *National Geographic* between 1970 and

1981. These articles created the background against which her own dramatic story would be told in later years. The two-article series of 1970–71 probably reflects Fossey's happiest days at Karisoke. Both articles are illustrated by Robert Campbell, a skilled and dedicated wildlife photographer who also became her lover for a time. The story from 1970 is about dispelling the King Kong stereotype: Fossey regales her audience with comic stories of Coco and Pucker and the heartbreak of losing them; the "wise old silverback" Rafiki and his attentiveness to the senile and dying matriarch Koko; the ascendancy of Uncle Bert, a "gouty headmaster," after the death of Whinny; and the unexpected playfulness of that same gouty headmaster when infants approached.[16]

The leitmotif in all of Fossey's work is her worry about poachers, shrinking habitat, and the looming extinction of her forest friends, but each of her articles is also punctuated with a few details about gorilla science or her own observational protocols. In the 1970 article, she describes acting the fool as she imitates gorilla gestures to gain their trust. In the 1971 article, Fossey continues her explanation of field techniques by describing the variety of gorilla sounds, which she also imitated to gain their trust and communicate specific feelings and intentions. The protagonist of this article is her friend Peanuts, a member of Group 8, whose appearances frame the story from beginning to end. In one of the accompanying photographs, Campbell manages to catch the iconic image of Peanuts just withdrawing his hand from that of a deliriously happy Fossey. The touching of hands, which recalls Jane Goodall's account of her similar encounter with David Graybeard, became a motif in popular primatology.[17] In this second article, National Geographic readers are also treated to follow-ups on the lives of some of the individuals introduced in 1970. Fossey's main concern here, though, is the gorillas' use of their range and habitat, especially for feeding—and the growing urgency of threats to the gorillas from habitat loss and poaching. She ends with an account of a violent poaching episode and worries that her study subjects might be a future target.

During the mid-1970s, as Jane Goodall had done before her, Fossey gained the scientific credentials she needed in order to continue the observations and conservation projects at Karisoke. Under the direction of Robert Hinde at Cambridge, she wrote a dissertation on gorilla behavior, and she produced academic articles on a number of topics relating to gorilla research, including vocalizations, ranges, diet, and female transfers between groups. She delivered professional papers at international conferences and contributed to an important anthology of research on the great apes. But from 1978 onward, she wrote urgently about Digit as an individual and a symbol of the perils faced by all mountain gorillas. The 1981 article in *National Geographic* is not about Digit, but it is partly a reaction to his death.

Fossey discovered that years of observing the gorillas revealed in them the dark side of primate being—not only the infighting connected with dominance hierarchies, which she expected, but also murder and mutilation, routine infanticides, and the possibility of cannibalism, which she could never prove nor disprove, although she diligently collected evidence. Goodall's similar discoveries about chimpanzees had temporarily sent her into a tailspin and permanently altered her feelings about chimpanzee nature. Not so Dian Fossey. Since she respected her predecessor, learned field techniques partly from a visit at Gombe Stream Reserve, and consulted Goodall's research for models and information, Fossey was prepared for the possibility that everyday gorilla behavior might include violence. What she eventually learned about their dark side fell short of the nervous rages and cold-blooded killings Goodall saw in her beloved chimpanzees, but the gorilla violence Fossey witnessed still might have shocked someone less prepared for it.

In the center of Fossey's 1981 *National Geographic* spread, a three-page sidebar with photographs and text by Karisoke student Peter Veit details the killing of Marchessa, the elderly matriarch of Group 5, by the young silverback Icarus. Icarus's antics had figured in Fossey's discussion of play behavior in 1970, when he was "a little wizened elf-eared fellow."[18] In 1981, however, he was a murderer, jumping on Marchessa's stomach; dragging her, while she moaned in pain, round and round the forest; punching her chest with his fists; and punishing her body for eighteen hours after she died. Veit's high-quality close-ups show Icarus dragging and sitting on the body, Marchessa's death grimace, and Icarus beating his chest in a display of power, with Marchessa slumped in the background. Marchessa's autopsy revealed parasitic cysts in a severely weakened body. In a view partially based on these findings, Veit speculates that Icarus's behavior expresses "puzzlement and frustration at her lack of response." Icarus also probably "knew Marchessa's death would enhance his power within the group" by ensuring that "Beethoven [the current reigning silverback] would lose a breeding partner [and] be pushed aside."[19] Neither Fossey nor Veit pulls punches: Fossey reprises this incident two years later in *Gorillas in the Mist*, adding that the youngsters and subadults in Group 5 participated in the abuse of Marchessa's body.

When a female is killed by natural or unnatural causes, her own and her future progeny's genes are lost to the gene pool. Male deaths are at least as significant. For example, when Digit was killed, Whinny lost power and was soon replaced by Nunkie, who killed both Whinny's and Digit's progeny. Sometimes the silverback is killed, as happened to Group 4's Uncle Bert, along with his mate Macho, as they defended their son Kweli from poachers. In the aftermath of that incident, Group 4 disintegrated further. "And so the Greek tragedy unfolded," Fossey writes. "Beetsme managed to kill Frito [another of Uncle Bert's offspring] only 22 days after Uncle

Bert's demise. It was Visoke's fourth known infanticide."[20] In spite of his efforts to lead the group, Beetsme simply lacked the experience and personality to lead, and the group fell apart.

Paradoxically, intra–gorilla group killings may sometimes have the effect of enhancing genetic diversity. Fossey found that infanticide, which figures into gorilla society more centrally than in chimpanzee society, is the norm when a silverback dies and a family group is taken over by a new silverback from a different bloodline. She speculates that in these circumstances infanticide can strengthen group cohesion and that the periodic replacement of one silverback by another potentially enhances the genetic health of the group because it diminishes the frequency of father-daughter inbreeding. Since the range of the mountain gorilla is so limited, maintaining genetic diversity is essential to the entire breeding population of the species, not just the group involved. Thus, in Fossey's view, if Icarus's behavior toward Marchessa does not betoken a gentle personality, it still represents a trait that is valuable in evolutionary terms. "Perhaps we will find," she concludes, "that the gorillas' own strategies of group growth and maintenance will circumvent group disintegration caused by man's encroachment."[21] Fossey therefore manages to make sense of a situation that puzzled Sarah Blaffer Hrdy in her study of langurs in India (discussed in chapter 4) at about the same time.

But Fossey also makes a compelling case that any human disruption of a gorilla group creates genetic stress that is likely to result in a domino effect of killings of adults and infants—if such actions are likely to support a silverback's power, family group solidarity, or the genetic diversity of the group. As the mountain gorillas' range shrinks, and when poachers kill individuals, group dynamics are unsettled, and transfers of power are liable to take place more frequently than normal. Time after time, human intruders have initiated a vicious cycle of killings, diminishing both gorilla population numbers and genetic diversity. If it was appalling for Fossey to witness internecine violence among the gorillas, it was more appalling for her to consider human responsibility for it. And unlike Goodall, Fossey seems to preserve, intact, the sense that animal behavior can suggest a better model for our own. The innocence of the gorillas was a matter of passionate belief for her, but this belief was in accordance with the strictest of scientific protocols, which mandate that only humans be judged by human yardsticks.

Ironically, her belief in the gorillas' innocence made it even harder to leave them.

IV

By the time she was killed in 1985, the world of the mountain gorillas had become Dian Fossey's whole world, and her murder belonged to the tragic patterns of that world. In an eerie repetition of Digit's death, she was found two days after Christmas, her skull split by her own panga before she could discharge the pistol in her hand. The Rwandan government charged one of her students and one of her former trackers with the murder. The tracker was said to have committed suicide in prison; the student left the country and was never extradited. No one close to these events believes that the men charged with Fossey's murder had anything to do with it. Both her critics and her defenders have tried to explain her death as a result of her politics and actions, but who killed her, and why, remains a mystery.

This much is true: Fossey was buried in the gorilla cemetery at Karisoke, next to Digit and in the company of many other gorillas who died during her time in the Virungas. Karisoke was her home and her castle, and in her writing she repeatedly refers to the Virungas as the gorillas' "last stronghold."

She defended this mountain fastness relentlessly. And if she referred to casualties within the gorilla population as "Greek tragedy," it was her own life that, at that point in history, could be so understood.

Fossey was unable to finish her own story, so others have finished it for her.[22] Driven by the flaw of single-minded protectiveness—or, as some of her detractors insist, pride— the character of Dian Fossey is transformed from the heroine of idyllic romance in Adrienne Rich's poem to a tragic hero in Farley Mowat's biography, as well as in the film that takes its plot from Mowat's book and its title from Fossey's. In the Kinyarwandan language, the locals called her "Nyiramacy-ibili," which she understood to mean "the old woman who lives in the mountains without a man," a derogatory name she claimed for herself. But it would be wise to remember that, in her exile, Fossey was not much different from many others who have walked the same path, such as her predecessors George Schaller and Jane Goodall, her student Amy Vedder, and her sympathetic biographer Farley Mowat. Fossey's single-minded pursuit of a life in the field has set a pattern for many others.

The setting in which Dian Fossey's life unfolded was a literal war zone, a tragically diminishing forest habitat, and a field created by human science and human dreams. Death is no doubt responsible for the fact that she became a media sensation, but she was also a dedicated scientist and conservationist. Even though she blurred the distinction between home and field, without her level-headed publications about the gorillas, even her spectacular death could have contributed nothing to the efforts made since to save them. Perhaps, before it is quite too late, human beings are learning from her death and her stories that another, even deeper tragedy would be the loss of our gorilla kin.

CHAPTER FOUR

Morphology of the Tale

We exist in a sea of powerful stories:
They are the condition of finite rationality and
personal and collective life histories. There is no
way out of stories; but . . . there are many possible
structures, not to mention contents, of narration.

—DONNA HARAWAY,
Modest_Witness@Second_Millennium

I

Taken together, *The Origin of Species* and *The Descent of Man* constitute not just a theory but a powerful story, which was accepted by most of the scientific community of Darwin's time partly because it was a satisfying narrative. When Darwin boarded the *Beagle*, he packed *Paradise Lost* in his suitcase, but by the time he wrote his great works, he no longer subscribed to the teleological assumption that human life is a progression toward a new Eden.

It is safe to say that Darwin's story has been credible, despite its outrageous affronts to religious sensibilities and human arrogance, because it is no mere chronological string of events: it is a plot, a linking of cause and effect, with a myriad of marvelous details presented in evidence. And Darwin's details do more than prove a point. They are story particles that add depth and intensity to narrative and provide aesthetic pleasure; conversely, it was from these story particles that Darwin was inspired to work out the story of life on earth.[1] Darwin often complained to his family that he found it difficult to avoid being dull or incomprehensible when laying out his theories in enough detail to persuade amateurs as well as experts, friends as well as professional rivals. But he knew what he was doing; he managed to persuade not by simply presenting evidence but by using it to add color to the plot.

For a scientist looking at the world in a new way, or a conservationist trying to save the planet one monkey at a time, or an environmentalist attempting to explain the unbreakable connection between a living creature and its habitat, the stakes in good storytelling could not be greater. Most of the primatologists whose work I investigate here see themselves as stakeholders in the struggle for a better future for primates of all kinds—and the rest of the living community as well. Most of them, consciously or not, are inspired storytellers. They need to be.

As long as science has existed, scientists have presented their findings in narrative form. Narrative is perhaps the most ubiquitous way of arranging information, because most human beings perceive and remember events in a linear way; E. O. Wilson, in fact, calls the human mind a "narrative machine."[2] But there are all kinds of narratives, and every narrative is inevitably selective. Some events must be left out, others emphasized. To "cut a long story short" is the craft of a good narrator.

Devices for shortening, selecting, and showing causal relations among events are often formulaic. Darwin lived before the day of the formalized scientific narrative, so he had to make up a form for his theory as well as present the evidence for it. The modern scientific monograph, however, follows a highly predictable pattern, partly to facilitate the verification and evaluation of reams of manuscripts by editors, and partly because—as my colleagues in the social and natural sciences tell me—the form makes scholarly writing easier. But these are not the only reasons the monograph is organized as it is. The sequence of scholarly writing in the sciences, in fact, reflects the sequence of the scientific method, as follows:

(1) Introduction: After much research and reading, the scientist realizes that an important question has not been fully answered or a difficult problem has not been solved. He (or she or they) makes an educated guess, or hypothesis, and the search for answers begins.

(2) Methodology: He (or she or they) reflects on how best to find accurate answers or solutions, through carefully recorded observations and/or controlled experiments, and makes a plan.

(3) Results: The scientist carries out the plan, in the laboratory or the field, and collects the desired data.

(4) Discussion and conclusions: Finally, the scientist sits down to make sense of the collected data. When all goes well (as it usually does, or the study would not be published), the question is answered or the problem is solved, at least partially.

The scientific article or book is an idealized abstract of the scientific process, and it is also something like a quest. The scientist is the hero of the tale; the search for answers is the quest; the answers or solutions are the Holy Grail or the princess, now rescued from the tower. Much is left out.

Most scientists conscientiously seek the truth; still, this formula could scarcely be more artificial. The form of the scientific publication suggests that the scientist proceeds in an orderly, linear fashion, from identifying a problem to finding a solution. In reality, the hypothesis may be revised many times to fit emerging patterns in the data. The original question may have already been answered before the study was undertaken. The conclusion may emerge temporarily from a matrix of doubt. The whole study may be motivated by dreams, desires, emotions, or ambitions. Finally, the scientist's personal experience, personality, and identity are submerged in the formal prose style; indeed, in many scholarly publications, even information that would reveal the researcher's biological sex is obfuscated by the substitution of initials for given names. The formalized grammar deprives the study animals of agency, ontological significance, (often) sex, and individuality.

II

Many primatologists rebel against the suppression of their identities and the suppression of details about their study animals. These rebels have urgent stories to tell. Certainly, primatologists are not the only scientists who write such stories, but as a group they seem driven to tell the whole truth, as they perceive it, about what they witness and experience in doing their work—the truth beyond what can be conveyed in an academic scientific publication. If the plots of field studies by Goodall and Fossey have elements in common with romance and tragedy, as the individual ape hero— human or otherwise—takes center stage, other primatologists have developed the genre of the autobiographical field study into a narrative art that in many ways coincides with the art of the novel.

This assertion that the autobiographical field narrative can be novelistic may seem strange. The novel is a long prose narrative, usually fictional, and usually about what *might be* true, rather than what can be verified or what has actually happened. There are only rare exceptions to this rule, such as the true crime novel or the historical novel that sticks closely to the factual record, and even in these cases, much of the narrative surrounding actual persons and events is speculative or fictional. Nevertheless, modern literary theorists suggest that what identifies a novel as a novel has more to do with its structure and rhetoric than whether it is fact or fiction.

One of the novel's distinctive features is what the anthropologist Clifford Geertz calls "thick description." The genealogy of this term is suggestive: it was invented by the mid-twentieth-century philosopher Gilbert Ryle, whose essays on language and arguments against Cartesian mind-body dualism have influenced generations of natural scientists, philosophers, literary theorists, linguists, and—through Geertz—cultural anthropologists. By "thick description," Ryle and Geertz mean the interpretive layer surrounding a bare chronological listing of events or behaviors. Such a listing would be tedious and devoid of meaning apart from what the reader or listener cares to bring to it, and so it cries out for interpretation by the author; indeed, it is difficult for the teller of the tale not to add an interpretive context. In addition to "cutting a long story short" through selection, the skilled storyteller, says Geertz, must be aware of the authorial drive toward thick description, use it, and control it. Not only that, but cultural anthropology demands the scientist's conscious construction of meanings that can be assessed, evaluated, and verified by others. Both the anthropologist and the primatologist (who may indeed be an anthropologist) make sense out of their material in the same way a fictional storyteller does.[3] As Hayden White points out in "The Historical Text as Literary Artifact," "it does not matter whether the world is conceived to be real or only imagined; the manner of making sense of it is the same."[4] Making sense of factual material involves creating the links that transform a chronology of events into a plot.

Whatever shape it takes, a plot is more than a "raw chronology." Rhetorical literary theorist Wayne Booth, who coined this phrase, counterposes raw chronology to the "story-as-told."[5] Literary theorists have developed other vocabularies to distinguish between a bare chronological listing and thick description: *fabula* versus *sjužet,* or *story* versus *discourse.* The writer of the academic scientific article or book relegates interpretation, or "thickness," to a discrete section at the end; the rest of the text could be said to be raw chronology. By contrast, the field narrative is a literary production, and, if not fictional, it nevertheless has fictional elements. It substitutes one kind of artifice for another; some persons and events are left out, others emphasized. And the interpretive layer pervades the whole narrative, not just the ending.

Thick description is not the only similarity between a field narrative and a novel. Not only is the raw chronology in both shot through with and surrounded by a halo of thick description, but events in both the novel and the field narrative take shape as plots, which consist of strings of causal relationships, dramatic suspense, and narratively satisfying resolutions. Writers create characters by endowing figures in their stories with individuality and interiority in a way that makes their actions plausible—and thus, to some extent, universal. Writers of fiction have at their disposal a number of techniques for creating suspenseful plots; some of the most common are constructing a frame story (a story bracketing another story), introducing flashbacks or flashforwards, telling the tale from multiple perspectives or with multiple generic markers, and weaving together two discrete but related storylines.

The writer of the field narrative may do all of these things without violating factual accuracy, but one, almost universal, strategy in the books considered here is the elaborate braiding of human lives with the lives of the study animals.

III

Monkeys and the humans who study them are typically characterized as social rather than heroically individual beings, and the most significant individual characteristics in monkey narratives are those that govern social interaction. There are several reasons why this should be so. All primates (even orangutans) are social, of course, and their societies abide by complex rules. Moreover, unlike King Kong or the all-too-real Digit, in Western literary tradition, monkeys do not contend against their surroundings with compelling and heroic force; except in rare cases, such as the Curious George stories or the Doctor Dolittle series (not coincidentally stories for children), monkeys do not set out alone in search of adventure. If apes have been characterized as large, powerful "half-men"—as distorting mirrors of human nature capable of terrorizing humans—monkeys have been seen as mirrors of human weakness and venality. In cultures ancient and modern, monkeys such as baboons and langurs are overdetermined by their coexistence with humans as pests, sacred animals, or both at the same time. In some modern narratives, monkeys are also overdetermined by their history as test subjects and, occasionally, working animals. All these dimensions of monkey existence have contributed to the ways in which monkeys are characterized in contemporary field narratives as social beings. Primatology narratives centering on monkeys tend to resemble novels rather than epics and romances; it is in novels that writers can best explore social interactions.

Primatologists tend to think of themselves as similar to their study animals. While they do not deny that they have significant human relationships, Schaller, Goodall, Fossey, and Galdikas appear in their own books as lonely individualists, braving danger and privation in their quests for knowledge. This self-characterization results partly from the circumstances of the field, of course. However, Goodall, Fossey, and Galdikas also convey a sense of strong, individual personalities in their study apes, who have the capacity for developing abiding cross-species relationships with their human observers. If Schaller does not go this far, he does describe the mountain gorillas as majestic and mysterious individual presences, and his own quest to find them as a solitary journey.

In contrast, the dangers faced by field primatologists who study monkeys are typically social and professional. Professional rivalries and misunderstandings can be perilous, but these problems are often described with a keen sense of the ridiculous. The "thickness" of narratives about monkeys has to do with the social complexity that the primatologist experiences and observes, and both monkeys and their human observers are more like characters from a nineteenth-century novel than a hero in an Arthurian legend. Dorothy Cheney and Robert Seyfarth admit as much about monkeys (and perhaps those who observe them) when they liken female vervets to "characters in a Jane Austen novel" who "attempt to maintain close bonds with kin, defer to those of higher rank, and simultaneously attempt to establish bonds with animals of high status." Later, they compare vervet family relationships with those described by Tolstoy.[6] Both of their books are replete with such offhand comments.

Novels typically consist of layered stories. In the dominant form of the eighteenth- and nineteenth-century novel (and the form of many novels written since), at least one plot centering on a group of working-class characters is interwoven with at least one story of an aristocratic family or social circle. Characters are dynamic individuals, and, at the same time, they are representative of their social class or way of life. Similarly, in Shirley Strum's 1987 book *Almost Human: A Journey into the World of Baboons*, the first part of the narrative is a collection of stories about individual olive baboons, woven into a story of the whole baboon community at Kekopey Ranch in southern Kenya; the climax comes when the entire troop narrowly avoids being exterminated after some baboons take to raiding the maize crops of the local farmers. Instead of two social classes, Strum focuses on two primate species. The baboon story is so perfectly balanced with the story of the scientists who study them and the farmers who live and work in the baboons' territory that, from the very beginning, the two plots can scarcely be disengaged from each other.

But any analysis of any tale must start somewhere, and this analysis will start with the monkeys. Strum's tale of the baboons begins with her attempt to habituate the Pumphouse Gang, so called in reference to Tom Wolfe's popular account of teen culture in La Jolla, California, and because the baboons live near an actual pump house. Strum's first day by herself at the study site coincides with the arrival of Ray, a large, handsome male baboon who likewise approaches the group as an outsider. Trained to expect male aggression as the driving force in a baboon group, she is surprised to see Ray observe from a distance for several days before making a move. Ray's first attempt to insinuate himself into the group is equally surprising: instead of rushing in to challenge the dominant males, he begins, carefully and tactfully, to follow Naomi, the lowest-ranking female, who always positions herself at the periphery, apparently to avoid conflicts with the other females. Ray persists for days. Naomi is cautious, though, and allows Ray to touch her only after his repeated gestures of friendship. Even so, Naomi's offspring remain aloof until she engages in a genuine, if nervous, grooming session with the newcomer.

Once his friendship with Naomi is established, Ray begins to climb the social ladder. Like all males in Strum's observations, Ray is a social climber, but he is bolder than most. Instead of seeking out the females ranking just above Naomi, he goes straight for Peggy, the highest-ranking female in the Pumphouse Gang. Calm and confident because her high status has always afforded her safety, Peggy offers less resistance than Naomi, but Ray's success with her is still not easy. Peggy's male friend Sumner becomes jealous. He threatens Ray continually, until, one day, the two males charge each other in such a frightening display that discretion becomes the better part of valor, and they part ways, leaving Peggy alone in the middle. Ray ultimately succeeds with Peggy and her numerous offspring, and as a result, he is quickly integrated into the troop.

Social status is not the only benefit Ray derives from this friendship. Like most other primates, olive baboons thrive on intimacy, and not just with individuals the same age. Peggy's children and grandchildren benefit from the family friendship with Ray by gaining extra protection and an extra play partner. They also confer protection on him (and all the adult males keeping company with them) because the presence of infants usually quells aggressive behavior among males. "Infant buffering," picking up an infant when someone else makes a threatening gesture, is, in Strum's observation, a common practice among adult baboons. (Other observers, including Robert Sapolsky, have posited less benign interpretations of this practice, but there is a growing consensus among baboon researchers that specific behaviors have different meanings in different populations—new evidence of baboons' behavioral flexibility.)

After this social conquest, Ray remains active and busy in the troop, sometimes challenging other males for place or copulation rights with a female, but Strum notes that the conflicts are quite often more ritual than real. Erect hair and the display of fierce canine teeth are usually for show; the canines of male baboons are so thin and fragile that they break easily, so even though biting someone else can inflict a lethal wound, the price is liable to be a broken tooth. Furthermore, Strum observes that winners in such conflicts don't automatically claim their rights when they win them. Often, a noncombatant male makes off with the estrous female while others are posturing, and sometimes the posturing alone apparently drains off interest, leaving the female free to pick and choose among various suitors or to take a recess from it all. Throughout the early part of his stay with the Pumphouse Gang, Ray challenges other males for dominance, once even trying with unmistakable gestures to enlist the aid of his human friend. Flattered, Strum resists Ray's blandishments and watches him win the bout on his own. Eventually, like most high-ranking males, Ray is confident enough of being accepted that he gives up fighting except on rare occasions.

From watching Ray and his associates over several years, Strum concludes that rank simply means not being displaced from a good food source, a grooming session, or a nice spot in the shade. Both males and females will usually give way instead of fighting, employing a variety of strategies to get what they want without aggression. Males move from their natal groups, and often after several years with the second group, they move again. (The evolutionary advantage to the group is avoidance of inbreeding.) New male members tend to be higher ranking than older males because the latter have developed social ties and strategies that make fighting and displays less useful. Females remain with their natal groups; their dominance rankings are stable and important, but asserting dominance is usually a low-key affair of eye flashes, gestures, grunts, or changes in location. Males have dominance over females because mature male baboons are almost twice the size of mature females; they can and do sometimes bully females to get food and other advantages. But the females have advantages of their own and often wheedle food and favors.

Ray and Naomi are among Strum's personal favorites, but she credits Peggy with teaching her more about baboon society than anyone else. Peggy's high rank within the troop probably derives from her mother's rank as alpha female in the same troop, combined with Peggy's own position as her mother's youngest surviving daughter—a pattern Strum sees repeated several times in this group and the others she observes over a period of many years. In any case, Peggy is "socially brilliant," and though large enough to bully the other females, she usually asserts herself with subtle gestures—and gets her way with the much larger males by guile, good timing, friendship with high-ranking individuals, and sometimes the aid of her eldest son.[7] Unlike Ray's dramatic adventures, Peggy's life follows a predictable pattern of domesticity: engaging in intimate exchanges with her family and best friend Constance; giving birth to Pebbles (and graciously allowing lower-ranking females to inspect the infant); losing her temper when a grandchild begs too persistently; and persuading her male friends to share meat. Crisis in Peggy's life is postponed until old age, when an illness provokes her eldest daughter, Thea, followed eventually by Peggy's other children and grandchildren, to abuse and ostracize the family matriarch for a time. When Peggy recovers from her infirmity, she never reclaims her alpha status from Thea, although Strum suggests that she could do so. Strum attributes this unusual episode in the life of the troop to Thea's nasty temperament and Peggy's natural grace. Peggy remains the second-ranking female in the troop until the end of her days. At the time she wrote *Almost Human,* one of Strum's most treasured possessions was the old matriarch's skull.

Ray and Peggy are individuals in this story. Strum not only explains what they do but also attributes interiority to them: they wish, hope, and plan. At the end of the story, when the baboons are captured and moved out of their home territory, one of them is so crafty that he avoids being caught and must be left behind; another watches intently, for several hours, for an opportunity to snatch an overlooked maize cob from an empty cage. Most of Strum's baboon characters are individualized, but these individuals also represent baboon nature in general—fluid (at least in comparison to most other monkeys), smart, adaptable, and "almost human," as Strum's anecdotes about them demonstrate over and over.

Like the characters in a nineteenth-century novel, individual baboons take advantage of the constant give-and-take within their social environments to advance their own interests and the interests of their families.

As a group, at the phenotypical and species level, baboons adapt to changes within the biological environment—a parallel to social and political changes chronicled in many triple-decker novels of the nineteenth century. During Strum's study, the males learn to hunt rabbits and small gazelles more effectively and cooperatively as changes in the surrounding countryside concentrate small game on Kekopey Ranch. Meat becomes not just a rare treat but a significant part of the baboons' protein diet, which before has consisted mostly of insects. And sharing food becomes more common as the diet shifts from small particles of food almost impossible to share—tiny onion grass corms, blades of grass, insects—to larger items such as fragments of an animal carcass. The changes that temporarily drive these prey animals into the Kekopey baboons' habitat, however, are related to larger transitions in the socioeconomic landscape around the ranch, which is eventually sold to the government to be partitioned among Kenyans moving from older lifeways or urban poverty into subsistence farming.

Thus, a sweeping change comes about in Kekopey baboon society: just as the new human residents convert to an agricultural life, so the baboons quickly learn to take advantage of newly planted maize crops. Almost immediately after the farmers arrive, the baboons start exploring their homesteads and playing with objects they find; in Strum's most comical anecdote, one of the juvenile baboons happens upon a large ball of bright yellow twine and proceeds to wrap a farmer's hut with it from top to bottom. Alas, most of their pranks are not so harmless. Easier to consume and higher in caloric value than their normal diet, maize enables the most aggressive grain-eating baboons to take longer rests and, eventually, to grow larger than baboons who stay away from the fields. Although all the baboons in the area sometimes eat the new food, the most incorrigible of the crop raiders split off from the Pumphouse Gang to form a new group aptly named "Wabaya" by the farmers and researchers. And "bad boys" (and girls) they remain, circumventing every effort made to discourage them: high fences, guard dogs, twenty-four-hour human patrols, even random lacing of the maize with chili pepper extract, a strategy that works well in Mexico against other kinds of crop raiders. Nothing deters them. Were their lives and the livelihoods of the farmers not at risk, the baboons' clever and indefatigable crop-raiding efforts would be hilarious. But the new habit is dangerous, and a crisis looms on the horizon.

The human plot in Strum's account is more twisted, and more is noticeably left out, including references to other scientists working in the area at the time. The human plot begins and ends not with cooperation but with professional rivalries and the sometimes oppressive power of hegemonic ideas. During this period in the discipline of primatology, olive baboons came to be considered the best model for the evolution of human behavior, partly because they are relatively large, sophisticated, and, like humans, savannah-dwelling generalists. Although chimpanzees offer a competing model for reconstructing early human evolution because they are genetically closer to *Homo sapiens*, Strum favors the baboon model because of their adaptations to savannah habitat. Baboons are, in fact, the most studied nonhuman primate— the subjects of research in the field, in zoos, and in laboratories—and the hegemonic idea when Strum began her research was that baboon society, like early human society, revolved around a male hierarchy established through aggression. Strum was in the vanguard of female primatologists who challenged this assumption.

One of the earliest texts in the development of this notion about male dominance, briefly referenced in Strum's story, is *The Soul of the Ape*, a field study of chacma baboons (not apes, of course) in the South African Transvaal by Eugène Marais, an early twentieth-century journalist, poet, and natural historian. Marais's experiences as a South African Boer living under British domination, while both the British and the Boers dominated the indigenous people, perhaps impressed upon him the human penchant for violence and dominance conflicts, especially among males. In any case, that is what he saw in his nonhuman study subjects. So committed was Marais to this view of primate nature that when he observed a large, gray-haired, nulliparous female taking part in fights and raids along with the males, he considered her a "monster" and had her shot so that he could perform an autopsy. Although, physiologically, she turned out to be perfectly normal, he expressed no remorse; Marais's view of primate nature was obviously heterosexist as well as sexist. For the most part, however, Marais did not notice individuals except as they struggled for status. He kept detailed notes on dominance struggles, sexual behavior, and addictive behavior, and suggested parallels between baboon and human evolution, but his account includes few stories of individuals or small groups within the larger community.

Soon after returning from the Transvaal, Marais published a series of newspaper articles about his experiences with the baboons there. He committed suicide a few years later, and the book he wrote about his baboon observations remained unpublished until 1969, when the novelist, playwright, and popularizer Robert Ardrey wrote a lengthy and enthusiastic introduction for the Penguin edition. Thanks in part to figures such as Marais and, much later, his champion Ardrey, the male dominance and aggression stereotype continued to dominate popular culture, as well as academic circles, from its first strong articulation by Solly Zuckerman in the 1930s, throughout the Cold War years, and into the 1970s. Finally, this model was successfully challenged by Strum and others. When a scientific theory so closely resembles conventional wisdom, challenging it is difficult.

The conflict that develops from the beginning of Strum's story is between believers in the centrality of male aggression and nonbelievers—or agnostics, as Strum herself was when, somewhat resentfully, she was packed off to Kenya by her professor Sherwood Washburn to study olive baboons instead of the more beautiful and rare patas monkeys she had hoped to investigate. A proponent of the idea that all primates are fundamentally violent and hierarchical, Washburn is an enigma in Strum's story: he never explains his insistence that she study baboons, but he also makes no attempt to undermine her scholarship, even when her conclusions begin to differ from his own. Without quite saying so, Strum suggests that Washburn was an unusually open-minded scholar, willing to be tested by his best students, as long as they rose to the challenges he set for them.

Such was not the case with all of Strum's associates, however. She recounts being treated by her male colleagues, on her first day at Kekopey Ranch, to an eye-opening jaunt in the project jeep to visit the Pumphouse Gang. Like most baboon scientists at the time, the other graduate students at Kekopey are addicted to the notion of male competition. They are so afraid of the animals they have come to study that they refuse to get out of the jeep—or even drive close enough for Strum to get a good look at the animals. Strum notes that she has changed some of the names in her story, and it's a good thing, too. Even at a distance, and during her first week, she looks more closely than her fellow student Matt: Dieter, one infant whom Matt has supposedly been watching for weeks, is actually a female, she discovers. The infant's name is quietly changed to Deirdre. Strum remains puzzled by this mistake: the infant male baboon grows to fit his penis, which at first is so disproportionately large that it looks like a "pink fifth leg."[8] Eager to leave Kenya after his grant expires, Matt tells scary stories about life in Africa and answers her questions with glib generalities, but his misidentification of Dieter and several other individuals convinces Strum that she can indeed study baboons with the best of them, despite her city-girl background. And if Matt's field methods are questionable, then the conventional understandings of baboon behavior, which he shares with almost everyone else, are also open to question.

In fact, about the time that Strum was making her way to southern Kenya, another way of studying baboon society had already been undertaken by Jeanne Altmann. Her research on baboon mothers and infants between 1971 and 1978 at Kenya's Amboseli National Park identified more to observe than just male dominance and aggression, and she proposed other interpretative methods. Predictably, until recently, most women primatologists were herded into the study of females and infants, but this gender segregation within the discipline has had an unexpected positive effect. The research relegated to women suggested, as Altmann says in the conclusion to her book *Baboon Mothers and Infants* (1980), a more "holistic" approach to understanding baboon—and human—society:

Any effort at isolating variables will always have to return to the questions: In what range of each variable does a human or other free-living animal usually find itself and with what combination of values of different variables is it usually faced? How do these variables interact? In the real world mothers and infants are not faced with one variable at a time. The effect of any variable is dependent on the values of others, often even in nonmonotonic and discontinuous ways. . . .

Despite the potential pitfalls in a holistic approach, there will always be limits and distortions to a reductionist or single-dimensional

world-view. We share with baboons a complex existence that is more
than the sum of its parts.[9]

Strum conducts her research holistically, getting out of the
jeep and walking and resting with the baboons from sunup to
sundown. A few years behind Altmann, she begins her research
by testing the dominance model for males and filling in some
of the blanks that remain with respect to the females, especially as females figure into the social hierarchies.

She finds that the overlooked female hierarchies are, in fact,
the foundation of baboon society because a female remains
with her natal group and inherits rank from her mother. Since
rank is inherited, violence among females usually amounts to
nothing more than occasional bickering over food and place;
at worst, it involves competition over handling new infants,
which occasionally ends with the kidnapping of an infant
belonging to a low-ranking mother. Strum notices that female
friendships, such as the bond between Peggy and Constance,
are stable over time and that children often inherit friendships from their mothers. These alliances provide safety nets
for infants, social intimacy, and support during conflicts.
Strum also observes that, as in human society, females tend to
form friendships with other females of similar rank.

Male behavior eventually interests Strum even more, not
because of fighting and jostling for power, but because olive
baboon males, she realizes right away from watching Ray,
have to develop a variety of social skills in order to make
friends when they transfer from their natal groups to new
ones. To her surprise, the most valuable friendships for males
are those they form with females, and, like Ray, they often
move as far up the social ladder as they can to ingratiate
themselves with high-ranking females. There are apparently
many advantages for male baboons who behave this way.
Certainly, friendships can provide mating opportunities, but
they also afford the social intimacy that baboons crave, which

is most often demonstrated by mutual grooming. The least expected benefit, male friendships with the infants of their female friends, turns out to be one of the most important.

In 1978, still a junior academic but back in California with six years' worth of data and revolutionary new findings, Strum arranged a symposium for eighteen baboon specialists, underwritten by the Wenner-Gren Foundation, a longtime supporter of primatology research. Her early papers from Kekopey, on predation, were well received because they were exciting but not controversial. This time, Strum recounts in *Almost Human*, when she tries to present her new model of baboon behavior, emphasizing negotiation and exchange of favors rather than dominance and aggression, she finds herself in "academic waves of capsizing dimensions."[10]

Most primate conflicts are solved in one of two ways: through violence, which is relatively rare, and through the conflict-resolution or avoidance techniques that Strum observes among the baboons. Baboons develop friendships partly in order to keep low-level conflicts from escalating into violence, and they literally turn their backs on individuals they wish to shun or fights they want to avoid. In baboon society, turning your back means doing without a snack. In subsistence human societies, shunning can amount to a passive death sentence. And in academic circles, shunning can result in a professional death sentence. That is almost what happens to Strum at the Wenner-Gren Conference. When she presents her conclusions from watching Ray, Peggy, the rest of the Pumphouse Gang, and the neighboring troops on the ranch, her work is met with outright hostility. Strum lists the accusations leveled against her: "I had invented my data, I didn't have enough information to draw the conclusions I had come to and that there *had* to be a male dominance hierarchy among Pumphouse males. I had managed to miss it, that was all."[11] But for the most part, her work is simply ignored; her colleagues turn their backs, just as a baboon does to avoid coming to someone else's aid, grooming a friend, or sharing food.

Strum comes away from the conference temporarily crushed, but with a new understanding that the culture of science, like any other culture, operates according to received wisdom and accepted rules. The dominance hierarchy in her academic discipline is fully operative. Finally, she reaches the conclusion that "scientific accounts of human social origins are the functional equivalent of myths."[12] Although she by no means advocates giving up science—in her view, the best hope for understanding human nature—Strum does insist that, as a culture, science needs to acknowledge the nature of its own narratives. What the scientist sees is in part determined by what he or she looks for, and that by what is already expected. As for her professional life, Strum survives because of the quiet support of a few "silverbacks" and a new interest in collaborating with Bruno Latour, a sociologist who studies the culture of science. A sadder and wiser woman, she also realizes the need for a new feminist science, as one way to question and revise received wisdom and to bring about the paradigm shifts described by the mid-twentieth-century giant in philosophy of science Thomas Kuhn in *The Structure of Scientific Revolutions*. For Kuhn, the stories of science are constrained by their historical context, and these stories can change in fundamental ways only when research converges with general acceptance within the field, as it did for Darwin's theories.

Layers upon layers. The human story in Strum's book is a weaving together of personal courage in the field and personal challenges in the world of academe. Her experience at the Wenner-Gren Conference is not unique. In Strum's jaundiced view of (human) primate society, the stereotypically human and the stereotypically simian are ironically reversed. Friendships and rivalries among baboons, which are transparent (to other baboons, anyway) and conducted with awareness of the ramifications of inflicting harm, are seemingly less treacherous than those among humans (or primatologists, at least). Not only do Strum's colleagues ignore evidence that conflicts with their own cherished beliefs, but graduate students steal from their professors and peers fail to acknowledge mutual findings, even when these findings could support each other.

Strum's narration of her experience within the profession resonates with that of Goodall and Fossey. And it is not transparent. Real-life characters are left out of this thickening plot, and names are changed. Although professional ethics may be a consideration, these omissions amount to Strum's own kind of shunning—and her "novel" is none the worse for it. Petty villainy is a general principle in the multifarious human world of this story, not so much a personal vice. The villains remain trifling, close-minded, and nameless. Any reader who is part of this world knows who they are in "real life." The rest of us are left to consider more important things.

Meanwhile, back at the ranch, the baboons and the farmers are in crisis: the raiding of maize crops has continued, and when Strum returns from California, one animal has already been shot. Despite Strum's attempts to remain observant but aloof, she must intervene in the baboons' lives in order to save them. If the turning point of the baboon plot is the introduction of maize farming and crop raiding, the crisis comes many months later with the transfer of the Pumphouse Gang and two additional troops from Kekopey Ranch to other ranches over a hundred miles to the north, near the Ndorobo and Samburu Reserves. Without the cooperation of the ranchers, the Kekopey baboons would be destroyed, and that would be a loss not simply of and to the animals themselves, but to the research community.

By the time of the removal, Strum has become acquainted with Jonah Western, an expert in long-range planning for African people, animals, and ecosystems. When the farmers first move into Kekopey, Strum sides with the baboons, whose way of life is compromised by the introduction of the maize crops. But with Western's help, she eventually understands what Goodall discovered after many years and Fossey may never have realized: the welfare of humans and other animals is connected, and one cannot take sides. Although she is just as invested as Fossey in her study animals, as acclimated to the landscape they inhabit, and as accustomed to having her social needs fulfilled by nonhuman primates, Strum still manages to separate herself from the animals in order to work more effectively for their survival. Her success in this effort may be partly explained by the flexibility of the baboons themselves, in striking contrast to Fossey's mountain gorillas, who were unable to adapt to new habitat and new social pressures.

It takes everyone involved to make the removal project work, and this part of Strum's book provides an interesting overview of the obstacles presented by Kenyan postcolonial politics and economics. Well versed in available resources and those who control them, Western contacts likely hosts for the soon-to-be-homeless baboons. One by one, Strum visits his selected locations. She looks at every spot with "baboon spectacles," noting abundance of wild foods, water sources, and good cliffs for sleeping. One by one, she eliminates potential homes for the baboons until, in a cliff-hanger, so to speak, she finally compromises, sending most of the baboons, including the Pumphouse Gang, to Chololo Ranch, owned by three brothers of the Saburu tribe, and the other, less-studied monkeys to a more remote location at Colcheccio Ranch, owned by an Italian expatriate. One group that has not engaged in crop raiding stays at Kekopey, where, everyone hopes, it will continue to refrain from eating maize and thus be able to coexist with the farmers.

Strum is triumphant, but the baboon tale is not quite over. The removal itself is not without risks. Since no baboon troop has ever been relocated in the wild before, she has no protocol to guide her efforts. However, the project proceeds with the help of Western, Strum's Kenyan research assistant Josiah Musau, a large trained staff composed mostly of local people, photographer Bob Campbell, human and monetary resources from the Institute of Primate Research in Nairobi, and Mary O'Brien, who has experience translocating deer in California.

The details are as exacting as the big picture. Finding enough appropriate cages for the baboons is the first problem. Baiting the cages in such a way that the most devious baboons are unable to retrieve the tempting maize cobs without tripping the doors is more complicated than the humans have figured. Luring the more cautious baboons into the cages after they have seen their adventurous fellows already trapped—another problem. Grouping cages of families and friends together and finding vehicles and drivers to move all the baboons at once so that their social life will not be disrupted proves to be the most complicated challenge of all. Then Strum and her associates have to calculate how much to provision the baboons once the troops are in their new homes, so that the animals will not suffer but will also quickly find and eat the new available foods. Adjusting to new sleeping quarters and finding the nearby watering holes could also be problems for the baboons.

Despite these challenges and her fears, Strum is grateful for the chance, at long last, to study the sedated baboons up close and to touch them, as she has longed to do since the beginning of her association with them. And, fortunately, the baboons in the Pumphouse Gang adjust to their new home in a matter of hours. They are willing to try the new foods, canny enough to find water within a day, and flexible enough to settle into new and different sleeping cliffs, although they do not accept the dormitory Strum has picked out for them. Families and friends reunite joyfully, and the social lives of the translocated troops remain intact.

Through all of this turmoil, the human plot continues. As in a Jane Austen novel, a wedding figures into the ending of Shirley Strum's story. After finding a new home for the baboons, Shirley and Jonah are married by none other than Louis Leakey's son Philip, a member of the Kenyan Parliament. Primatology is a small world, here replete with storybook symmetry: births, deaths, adventures, conflicts and friendships, trials and tribulations, and finally happy endings for everyone, baboon and human alike.

Scientists have not ultimately agreed that the baboon model is best for theorizing about human nature and human evolution, but Strum insists that it is. In *Almost Human*, the suspenseful plot in itself suggests a more compelling argument for this position than the scholarly foundation of Strum's work. It is a whopping good story with a hook at the beginning, adventures and suspense in abundance, a happy ending for all the good guys, and backs turned, finally, on the bad ones. The coherence of the story emphasizes the coherence of Strum's views and reflects her confidence in professing them.

This is not to say that Strum has not changed. The battering Wenner-Gren Conference in 1978 seems to have been a watershed in her life, and perhaps in the discipline of primatology as well. Strum herself realized that science studies—the critical analysis of scientific practices and assumptions—must be incorporated into primatology in order for the field to thrive and contribute to the ways in which human beings understand organic nature, including themselves. Life means change. Novels—and autobiographical field narratives like this one—teach that.

IV

Between 1971 and 1975, the anthropologist Sarah Blaffer Hrdy studied six troops of Hanuman langurs at Mount Abu in northern India. During this time, she also made briefer trips to other parts of the country, where she visited other langur troops and for comparison observed other colobines, a subfamily that includes colobus and proboscis monkeys as well as langurs. Hrdy's field narrative about this experience, *The Langurs of Abu*, tells the story of the puzzling and conflictive relations in langur society, including those between male and female langurs—a war between the sexes carried out through violence and deception. It is not a happy story.[13]

If the thickness of some plots comes from layered stories, such as the human and baboon plots in *Almost Human*, the thickness of others derives from layered generic conventions. According to Mikhail Bakhtin, the novel is a free space that allows for the introduction of many different literary and nonliterary forms.[14] Poems, advertisements, diaries, plays, and libretti are likely to be embedded in novels, and some novels consist entirely of letters; contemporary novels may include e-mail messages and excerpts from blogs. In Bakhtin's view, this generic flexibility, or "heterology," is a distinctive feature of the novel. *The Langurs of Abu* weaves together epic, soap opera, high drama, murder mystery, and even jigsaw puzzle—a mixture that challenges the very concept of scientific publication. In Bakhtin's theory, the novel is not only heterologous but "intertextual": it typically includes references to other tales, other times, and worldviews besides those of the narrator or the novelist. Hrdy's field narrative is richly intertextual, and this intertextuality is one strategy by which she problematizes human life, primate evolution, and, especially, the significance of sexual dimorphism and sex-linked behavior in primates.

The Hanuman langur, *Semnopithecus entellus*, is named for a god. "Langur" refers to the monkey's very long tail. Hanuman langurs eat large quantities of foliage, which their stomachs are particularly adapted to digest, but they take fruit, roots, grains, pulse, and bits of animal protein when they can get them. Like most monkeys, they live in large groups, but their society, as Hrdy describes it, is quite different from that of the olive baboons, most obviously because each group has one dominant male and perhaps a few adolescent male hangers-on; surplus males tend to congregate in smaller all-male cohorts. In this way, they are more like Hamadryas baboons, who live in the mountains of Arabia and the Horn of Africa.

Hanuman is the name of a monkey deity, a hero of the Hindu epic *Ramayana*. In this story, Hanuman seeks out (the human) King Rama's wife, Sita, after she has been kidnapped by Ravana, the evil king of the underworld. Ravana has taken her to his palace in Sri Lanka, where he hopes to seduce her (a more interesting challenge than taking her outright). First, Hanuman has to overcome Sita's suspicions in order to lay plans for a rescue by King Rama; then he has to escape from Ravana's henchmen, who punish him by tying burning brands to his tail. The punishment backfires when Hanuman makes his escape by running across the island, setting fires wherever he goes, before flying back to the mainland. Finally, Hanuman guides Rama back to Ravana's palace with an army of monkeys to rescue Sita. Hanuman is always brave, clever, kind, and entertaining, but in this myth of transactions among males, Sita is the pawn traded back and forth for sex, status, fame, and favor.

The langurs' story, then, has epic and sexist cultural roots. Langurs are part of the Indian social landscape and members of the human community; they are tolerated, humored, and even revered. Indians feed them as casually as people in other countries feed pigeons in parks, and many of these monkeys are protected in temples. But they make pests of themselves, ruining tile roofs, raiding crops and gardens, and stealing food from shopping bags.[15] Hrdy recounts watching a langur reach through an open window at a bazaar and start a fire by spilling a can of kerosene onto a brazier. "I am not a true naturalist," she admits, for *if it were not for the fact that langurs interact as individuals, I could never have sustained my interest in them for five years. It was the high drama of their lives, the next episode of the colobine soap opera that got me out of bed in the morning and kept me out under the Indian sun, tramping about their haunts for eleven hours at a stretch. During the five seasons of this study, political histories of the Bazaar, Toad Rock, and Hillside troops were marked by dramatic power struggles between males, accompanied by more continuous but less obvious fluctuations in the relations between females and by the alternating antagonism and coopera-tion between these animals.*[16]

In Hrdy's story, the monkeys' overdetermined epic past is never far from present reality.

She goes on to introduce "the protagonists in this chroni-cle"[17]: the alpha males Harelip, Splitear, and Shifty Leftless; the high-status females Pawlet, Mopsa, and Mole; and lonely old Sol, the most heroic defender of infants and other females. These individuals do not become particular friends of the observer, as David Graybeard is a friend of Goodall, or Digit of Fossey. Hrdy never comes to love them as Strum loves Naomi's little daughter Robin, who once grooms the fabric of her shirt. Friendship, at least with humans, does not seem to be in the behavioral repertoire of langurs, and Hrdy is prob-ably too taken aback by the langurs' treatment of one another to care.

Rather, Hrdy finds these monkeys remarkable for their suc-cesses within the complex webs of langur society and the opaque political plots that she, as the hard-boiled detective in the tale, must decipher. Questions abound. "The study of free-ranging monkeys is like a jig-saw puzzle poured out from its box," she explains. "The cover picture and a number of pieces have been lost. The significance of isolated bits becomes appar-ent only after much of the puzzle has been fitted together;

some pieces remain mysteries."[18] How did langurs evolve to be as flexible as they are in diet, habitat, troop size, and behavior? Why do some troops stay in one place, while others are nomadic within their range? Why did one elderly monkey seemingly commit suicide by climbing down a tree to be eaten by a leopard? Grooming is essential for individual well-being and social cohesion, but why did two langurs spend half an hour grooming a neighborhood goat? Why don't rank (indicated by feeding order) and leadership (indicated by who makes decisions about troop movement) coincide? Why do the old females become warriors in defense of their groups, even though they typically lose rank and often go without food while others eat? How did Pawlet become the alpha female in her troop, fall to the bottom of the pecking order, and then regain her rank?

Initially, Hrdy is most curious about langur infant sharing, or allomothering. Langur females are powerfully attracted to infants and start taking them from the mother as soon after parturition as she will allow. Why do some mothers readily give up their newborns and others refuse? How do some adult females gain the authority to compel others to hand over their infants? Why are nulliparous females more interested in allomothering than multiparas? Why would a multipara insist on taking an infant, only to discard it minutes later? Some allomothers turn the infant upside down, fling it about, and quickly tire of it; since the infant's only survival strategies are whining and clinging to the nearest furry chest, anyone wanting to get rid of it has to find someone else to take it, pry it off by scraping or swatting, or even sit on it. Yet other allomothers are entirely solicitous. Why do some allomothers routinely treat their charges roughly, while others do not? Why are some infants more "popular" than others? Who benefits from these exchanges?

Eventually, Hrdy decides that the complex web of interactions between male and female langurs is even more striking and puzzling than the phenomenon of allomothering. Why would an alpha male in sole possession of a harem leave his troop to risk copulating with females from a neighboring troop? Why do females solicit outside males? Why do they fake estrus? Why are copulating couples routinely harassed by juveniles and adults of both sexes? Why do langur males routinely kill the infants after they take over a harem from another male? Why do langur females who have lost infants in this way react by soliciting the new alpha male? Most of all, why do the females tolerate this behavior when, by cooperating, they could easily defeat infanticidal males? (Similar patterns of infanticide have been observed in other primate species—Fossey's mountain gorillas are only one example— and are plausibly explained as protection or expansion of an individual's genetic investment.)

Hrdy never solves the riddle of the suicide by leopard or the cross-species grooming, nor does she make sense of Pawlet's startling career. She leaves to other observers most questions about how the monkeys use the resources of their home ranges. But her narrative offers at least tentative answers to the compelling questions surrounding allomothering and infanticide. Like the hard-boiled detective she emulates, Hrdy uses the data collected during her five-year longitudinal study to investigate the ways in which key behaviors, which at first seem unrelated, may be linked. She suggests that male competition for access to harems, female competition for rank within the bisexual troop (that is, composed of both males and females), troop defense by older females, allomothering, "sexual harassment," and male infanticides are all part of a complex of reproductive behaviors that evolved separately in males and females, whose reproductive strategies are in conflict with each other. In Hrdy's view, langur society is an unrelenting battle of the sexes.

Seven years after beginning her observations, Hrdy comes to understand that males and females probably "commit adultery" on the sly because exogamy increases genetic diversity;

furthermore, a female's langur's adultery may give the alpha male's rival at least the appearance of a paternal stake in her offspring and thus a measure of protection if political structures shift. Allomothering benefits juvenile females by enabling them to practice maternal skills, and it benefits multiparas by enabling them to assert dominance over other females and by giving them information about the sex and health of new troop members. Allomothering also benefits the biological mother by giving her a rest now and then, and it saves her the trouble of resisting other females who want to borrow her infant. Allomothering benefits the infant not at all, but the unusual richness of langur milk may compensate for the infant's time away from the mother's nipple. (More recently, Hrdy has argued that allomothering is generally adaptive in other primates, including humans, for infants as well as adults.)[19] Infanticide and sexual harassment, too, are adaptive, and they serve similar ends: successfully interfering with the reproduction of another individual may free social and material resources for one's own progeny. Killing the progeny of a rival male triggers estrous behavior in the troop's females, and the new alpha has a better chance of passing on his genes. Even the elderly female warriors are selfish, to use Dawkins's term: no longer in their own reproductive prime, they can still benefit reproductively from defending the progeny of sisters and cousins, with whom they share half or more of their genes.

The way Hrdy sees it, life is hard in Hanuman's tribe. Langurs have a limited-resource mentality, and sex is a zero-sum game. But she does not arrive at these conclusions solely on the basis of the statistics she has so diligently collected, the literature she has so carefully studied, or even a combination of scholarship and number crunching. Before she published *The Langurs of Abu* in 1977, Hrdy had already made most of this information available in scholarly papers and articles. But to solve the mysteries of this data to her own satisfaction, she had to construct a plot: "Compromising between pure speculation and precise records, the primatologist must summarize the lifeways of species," she observes.[20] As she says in the introduction, Hrdy means for that plot to be an economic story of costs and benefits; the langurs conduct themselves as miniature, fumbling Milton Friedmans, hoarding resources and hoping that benefits will trickle down if, individually, the monkeys maximize advantages and cut costs.

But this economic theme is not the whole story. Hrdy's plot is complicated by its thickness—a web of allusions to other plots that destabilize it and underscore its place in the literature of postmodernism. The matrix of her tale, compounded from the most violent master narratives of human civilization, is evident from the beginning. In the acknowledgments, she writes of the story's human characters in epic terms: her associates are Olympian, her husband, David Hrdy, is Odysseus, and the artist Virginia Savage is Circe. The myth of the epic hero Hanuman is summarized in the caption to the first illustration. Moreover, the story proper begins with a reference to Shakespeare's vicious tragedy *Titus Andronicus*, in which infants become pawns in the dangerous games of their homicidal, adulterous, unintentionally cannibalistic, and power-mad parents. "If more primatologists had seen this play before going off into the field," Hrdy remarks, "they might better have understood the behavior unfolding before them in the savannas and forests where monkeys are studied."[21]

If Shakespeare's crudest play inspires the outline for the infanticide plot, the late nineteenth-century naturalist playwright August Strindberg suggests the solution to another big problem in langur studies. Female langurs steal copulations from males in competition with the alphas of their own groups, present when they are not ovulating, and, with rare exceptions, refuse to stand in solidarity with their embattled sisters, even when they could do so without immediate risk to themselves. Why? Quoted at the beginning of the last chapter in *The Langurs of Abu*, an exchange from Strindberg's *The Father* parallels Hrdy's explanation of female langur reproductive strategies, which hinge on their ability to deceive the males: "You could give me a raw potato and make me think it was a peach," the Captain remarks to Laura. "Just one thing more—a fact. Do you hate me?"[22] Hrdy concludes that female langurs very likely do, or should, "hate" the males, but they hide it.

Ironically, female deviousness may not be their best defense, just as it is inadequate for the human women in Strindberg's plays. As Hrdy concludes, "For generations, langur females have possessed the means to control their own destinies; caught in an evolutionary trap, they have never been able to use them."[23]

V

When a small baboon stretches her lips across her teeth and grins, she is signaling that what she is about to do—jump on top of a playmate, slide down the back of a large adult male, or wrestle with her mother—is all in fun. Her play face frames the actions that are to follow with her own meaning and intention. The playmate, adult male friend, or mother—whoever will be on the receiving end of the play violence or practical joke—is given instructions about how to interpret her behavior.

According to the sociologist Erving Goffman, all human actions except "natural" or unstudied physical or symbolic activity are framed, and framing is so ubiquitous as to be an unconscious, if necessary, part of human social life. So it is for other primates and for primatologists. Hrdy's research presented extraordinary challenges. If her conclusions are framed in terms of economic and political history, these human enterprises are only one of many frameworks invoked in *The Langurs of Abu*. Because the langurs she studied were integrated by custom and tradition into the daily lives of the human community where they lived, Hrdy also frames her research within the story of the monkey god for whom they are named. But these langurs do not, after all, resemble Hanuman very much; if they represent male dominance and a certain level of craftiness, they are not particularly valorous or clever. One cannot correctly read the langurs' story in terms of epic or myth.

In addition, Hrdy frames their social interactions as a mystery story and herself as detective, and the behaviors of the langurs remain mysterious, despite her efforts at detection. Ultimately, the answer to the question of why infanticidal reproductive strategies have evolved is indeterminate—the dark spot under the lamp. Darwin helps Hrdy explain how these strategies evolved, but motive is in part a metaphysical question, apparently irresistible to detectives in mystery stories but often left unanswered, even when killers, victims, and means are revealed.

Hrdy most consistently and successfully frames her research in dramatic terms: the langurs' daily lives are a "soap opera," occasionally marked by "high drama." When she compares female and male langur reproductive strategies to the plays of Shakespeare and Strindberg, she finds the language she is searching for in literary representations of gendered violence in human social life. Strindberg's work participated in the naturalist movement of the late nineteenth century, which was directly influenced by Darwin's account of biologically inherited behavior. The naturalists often referred to their fictions as "laboratories"—much like the "thought experiments" of philosophy—for working out what humans are likely to do in life-or-death situations. Not surprisingly, human life in these novels and plays is almost always represented as "nasty, brutish, and short." Citing Strindberg in the langur study enables Hrdy to subtly suggest that her research functions as both field study and laboratory science, although most primatology field narratives are framed as something other than science. It is difficult to imagine a study more situated within conflicting frames than this book—or, indeed, more intractable subject matter than the lives of langurs.

Hrdy's own readings of langur reproduction have evolved somewhat over time, as she has incorporated her early scholarship into more general primate behavioral and social theories, which include humans. *The Woman That Never Evolved* (1981) critiques what is perhaps a feminist myth of female cooperation or sisterhood and argues that human female competition is often carried out in behaviors, signs, and language too subtle to be reduced to the statistics that enabled Hrdy to reach conclusions about langurs. *Mother Nature* (1999) is Hrdy's attempt to explain human motherhood as a web of psychological, social, and cultural factors, combined inextricably with the physiological and endocrinological factors that come to us from the genes we share with our primate kin.

Hrdy has continued to rely on master narratives as frames for the conclusions she draws from a vast and increasing store of knowledge. Both of her later books are replete with stories and story particles from beyond the boundaries of science. And, for Hrdy, human behavior remains a story of games, players, and rules, of mystery and intrigue, of murders, motives, and means. Her best strategies for articulating the problems and solving the riddles of human behavior remain the "old themes that our fabulously inventive, and devious, species creates daily."[24] She elaborates: "The Empress Livia, Lady Macbeth, Mrs. John Dashwood [a character from Austen's *Sense and Sensibility*], and Strindberg's two genteel ladies who ever so subtly put one another down while conversing sedately in a Stockholm café—these happen to be my favorite examples. But returning now to science: where is the hard evidence for these competing ladies? Without it, we remain in the realm of fiction and opinion, and this world does not lack the critics who will promptly tell us so."[25] Regardless of whether there is hard evidence for her claims about male violence and female deceitfulness, Hrdy's stories serve as compelling feminist explanations for seemingly intractable human problems. And she has continued to follow these threads, discovered early in her career, in ever more powerful narratives about humans and their closest kin.

Thelma Rowell, a specialist in baboons and other Old World monkeys, remarked in a review for *Natural History* magazine that Shirley Strum's *Almost Human* "is an honest, in some ways an outrageous book because it treats together things that happen together that convention keeps apart: science and politics (in the widest sense) and the life of the scientist."[26] This observation about Strum's narrative approach applies almost as well to Hrdy's books, which treat together economics and biology, gender politics and behavioral observation. Rowell's comment also resonates with Bakhtin's definition of the novel. "After all," writes Bakhtin, "the boundaries between fiction and nonfiction, between literature and nonliterature and so forth are not laid up in heaven."[27] What turns the scientific article or book into the novel-like field narrative is the thick description, wherein lie the details that convention leaves out of the scientific account: observed primate behaviors that cannot be codified, the politics of the scientist's findings, and the details of the scientist's life. In other words, all that surrounds and influences science. Outrageous indeed!

Rowell's comment reflects a moment in history—1987, the publication date of *Almost Human*. Strum's narrative itself reflects the moment of its creation: it resonates with Fossey's book and foreshadows a more widespread habitat crisis. In both Strum's narrative and Hrdy's, the novelistic form suggests the indeterminacy and turbulence of this period in society at large, within the discipline of primatology, and in the natural environment, increasingly under pressure as the interests of humans and other animals collide and as new knowledge conflicts with received wisdom.

CHAPTER FIVE

Primate Characters

The prospect of using communication as a window on the
feelings and thoughts of animals seems the most promising if
only because it is so useful with our own companions.
In one sense animals may already be using the window, as
they succeed in conveying to one another their feelings and
simple thoughts. If other animals can get these messages,
cognitive ethologists with the advantage of the human
brain should be able to do as well.

—DONALD R. GRIFFIN, *Animal Thinking*

I

Every nonfiction writer, including every author of a primatology field narrative, has to answer two questions. The first is
epistemological: How do I know what I know? The second is
rhetorical: How do I convince others of my credibility? The
author has to discover the meaning of primate behavior and
convince others—scientists, environmentalists, and armchair
adventurers—that her information is correct. She must portray primate characters, including herself and her study animals, in a way that is narratively satisfying and at the same
time scientifically plausible. That is the heart of the matter.

In order to convey character, both Karen Strier and Barbara Smuts avail themselves of unconventional rhetorical strategies within conventionally structured scientific field accounts. Although their books are atypical within the genre I explore in this study, their strategies of primate characterization demand attention, partly because they are successful and subtle reactions to the internal pressures within the enterprise of primatology. In contrast, Biruté Galdikas's orangutan narrative *Reflections of Eden* is unabashedly about primate personality. Galdikas was the third of Leakey's "trimates." Because her way was paved by Goodall and Fossey, chimpanzees and gorillas, she was at greater liberty to portray her study animals as developing characters.

Portraying nonhuman subjects as characters is in many ways hostile to the goals and protocols of Western science, but scientists participate in their own special culture in addition to the larger culture of art and story that surrounds them. Primatologists, especially, are often pulled in two ways at once by strong cultural currents, which can be described as the personal and the professional, empiricism and imagination, or the objective and the subjective. Writing a successful narrative for the scientific community requires that the scientist negotiate around this internal conflict; writing a successful narrative for a lay audience requires that the scientist explain nonhuman animals in human terms that are sympathetic and understandable.

Although the discipline of primatology has incorporated the findings of Goodall and Fossey and validates the field methods that produced their results, tensions between objectivity and subjectivity remain in the practice of primatology. Science dictates objectivity, and within the strictest protocols of scientific writing—the poster, conference paper, or published article or book—the ape, monkey, or prosimian is "subject" in only one sense. It is not a being with an independent life, but an "object" of study whose being and behavior can be described as data that have been rigorously recorded and are verifiable, depending on the credentials of the observer.

At the core of every behavioral study is the ethogram, a list of specific behaviors compiled at the beginning of the study. The researcher counts these behaviors by ticking off incidences on a prepared form or in a notebook or tracking them with the aid of software developed for this purpose. There are good reasons for such an approach: since 1898, when E. L. Thorndike published a foundational monograph on researching animal intelligence, scientists have noted that, at least to a degree, carefully recording the researcher's methodology and the animal behaviors under consideration protects the researcher from working backward from desire and supposition to unfounded conclusions. When this happens, the scandal within the scientific community quickly spreads to the educated world at large. Anthropomorphism—seeing human traits in the animal subject—can lead to incorrect conclusions and can even indicate arrogant anthropocentrism. In many cases, critical distance from the study animal is also essential in order to obtain good data. Even in quantum physics, it has been understood for a decade or so that observing a phenomenon—say, photon movement—changes the phenomenon. So how much more does a living, intelligent being—a lemur or a golden lion tamarin, for instance—respond to the observations of a human researcher!

For birds or bats or bees, the ethogram functions in a relatively straightforward way, although not always perfectly. But apes and monkeys elude such attempts to package them. Their lives and personalities exceed the data that can be recorded in ethograms and morphological descriptions. Apes and monkeys are so like us in behavior and physiology that outside the most rigorous boundaries of the professional publication (and sometimes even within them), it seems impossible to represent nonhuman primates without suggesting individuality, personality, knowing, agency, emotion, and ontological value. And it seems to be impossible to live in the company of these animals for any length of time without feeling a personal relationship with them as individuals and as a social group. Ever since Goodall insisted that the chimpanzees of Gombe demonstrated individuality and psychological flexibility, most primatologists have assumed that nonhuman primates have an internal life. But an internal life is not quite the same as a theory of mind, so the question is still vexed and continues to be debated at conferences and in scientific publications. This sticking point may account for the reluctance of scientists to use the word "personality" in scientific discourse. Primate behaviorists still rely heavily on the ethogram because it is a useful tool, but it is always an overflowing container. The overflow finds its way into the tales told at parties, bursts from the interstices of scientific papers, and inspires field narratives.

The frustrations of the primatology enterprise—"real" work versus the overflow of feeling—are revealed indirectly and directly in the way apes and monkeys are represented. Goodall pioneered the practice of naming study animals in the field rather than numbering them, with matrilines designated by the same initial letter. In and of itself, the new practice of naming brought primatology closer to allowing anecdotal evidence and made intuition more respectable; naming confers individuality and personality and suggests an animal's capacity for intersubjective relationships with others. Goodall herself pointed out that naming has a second, entirely practical advantage: names are easier to remember than numbers. (Nevertheless, this development is not universally regarded as a good thing, nor is it universally practiced now.)

Characterization of nonhuman animals is riskier business than describing other humans. In *Almost Human*, Strum complains that *National Geographic* falsified a photograph of baboons peering through a window at her Siamese cat by erasing the unseemly splash of monkey urine under the windowsill.[1] Several years after her startling early discoveries about chimpanzee tool use, hunting, and highly individualized personalities, Goodall agonized about how (and whether) to represent darker truths of individual and species behavior. In *Gorillas in the Mist*, Fossey writes about the dilemma of representing Digit after his death: Should he be allowed the dignity of privacy and a quiet burial, as befitted his personality, or should the image of his headless body be used to raise funds for gorilla protection? Cleaning up the image of the baboons suppressed information; at the very least, the urine stain would have informed *National Geographic* readers that curiosity about unfamiliar animals does not inhibit a particular natural baboon behavior, although the photo is potentially less offensive if the stain is gone. Telling the general public the truth about warring, cannibalistic chimpanzees, as Goodall does in *In the Shadow of Man* and other publications, could have diminished the political will to protect these animals and undermined efforts to raise funds for their well-being. In Fossey's view, at least, the pathos of Digit's death could have placed the gorillas at greater risk because it encouraged more animal lovers to visit the Virungas, bringing new dangers to the gorilla families who lived there. All of these instances reveal stresses in primatologists' constructions of their nonhuman characters.

The primatologist has to be concerned not only with the way she reports or represents her study animals but also with the real limits of her own perceptions and assumptions. Representing the interior life of another person, human or otherwise, is always risky because we can never truly know their "unique experiences." The notion that all we can ever know with absolute assurance about another person is the sum of that person's actions was broached by the eighteenth-century Enlightenment philosopher David Hume, who discovered in this view a kind of cheerful freedom. The notion was picked up again by the gloomier Danish theologian Søren Kierkegaard at the end of the nineteenth century and further developed by the war-weary French existentialist philosophers of the mid-twentieth century. Objectivity in science and modernist literature and art became a good in itself. It is not an accident of history that the protocols of modern science in the West—also an outgrowth of the Enlightenment—coincide with this existential view. Jean-Paul Sartre, in fact, insisted that describing the mind and heart of a character, even in fiction, necessarily undermines plausibility and violates a covenant between author and reader. Needless to say, the difficulties are multiplied when the subject is not human and the genre is not (usually) considered literary.

Literary tastes have changed since 1947, when Sartre published *What Is Literature?* Literary theory has embraced, in many different ways, the problem of representing characters whose experiences cannot be directly known. Paradoxically, one of these strategies has been to displace human language from the center of human experience. For contemporary cultural theorists and the postmodernist cultural workers they study, mental activity and communication do not always depend on words, and despite vigorous objections, especially from the field of linguistics, the definition of language itself is under renovation. Scientists, too, have observed that complete objectivity is difficult to achieve and that intuition plays a part in discovery and understanding. For a gathering of scientists and science analysts in Brazil in the late 1990s—a meeting that ultimately led to *Primate Encounters*, the dense "state of the science" volume edited by Shirley Strum and Linda Fedigan—such epistemological questions gravitated to the center of their dialogue, even though the original purpose of the meeting was to assess changes in ideas about primate society.

If nonhuman primates could only talk! We would be on more solid ground, then, wouldn't we? The desire to understand and represent them accurately drove early attempts to teach apes (and dolphins) to talk. Soon it became clear that the anatomy of their mouths and throats is so unlike ours that they cannot generate human speech, so these attempts were replaced by instruction in sign language for apes such as Koko, with whom well-known language experiments were conducted in the 1970s at Stanford. The development of ape sign language was followed by the invention of a system of lexigrams taught to chimpanzees and bonobos—most famously the bonobo Kanzi—at the Yerkes National Primate Research Center outside Atlanta (the apes now live in Iowa). These apes learned to manipulate computers or portable signboards to communicate with humans and other apes in their linguistic community.[2] At the same time, science fiction writers almost too numerous to mention—some inspired by Robert Yerkes himself—have fantasized about apes and monkeys on whom cyborg implants or genetic manipulations have conferred the power of human language. (Physicist David Brin develops this fantasy in the monumental six-volume Uplift series; the favorite byword of his speaking chimps is "By Goodall!" and a central location in one novel is Mount Fossey.)

Most primatologists are good at imitating the sounds made by their study animals, and quite a few fiction writers (including, of course, Hugh Lofting) have speculated about learning animal languages. But in our dealings with other species, despite recent research into gesture communication, we humans have mostly remained true to our history as political and linguistic conquerors. We want animals to speak to us in our own language, and the fantasy remains that if we are successful at this, we can reach across the species barrier to represent them more truthfully.

Be that as it may, talking to animals in a laboratory reveals in only a limited way how free-living nonhuman primates experience their lives, and researchers still have to rely on an unsettling combination of data and intuition for conclusions about the meaning of simian behavior. (Unfortunately, the fad of language experiments in primatology also resulted in the scandalous abandonment of numerous signing apes to biomedical labs, warehoused and managed by technicians discouraged from communicating with them. Many are finally being consigned to sanctuaries, where they can be socialized and rehabilitated.) Clearly, for the primatologists represented in this study, apes, monkeys, and lemurs *do* talk. The trick is to hear them.

II

Muriqui monkeys (fig. 1), or woolly spider monkeys, are a critically endangered species now confined to a few fragments of the once-extensive Brazilian Atlantic Forest. These large, open-faced, golden-gray, soft-furred primates usually live in peaceable groups of males and females. Males are not domi-nant and do not fight for females, who are so receptive when they are in estrus that sometimes the males can simply line up and wait their turn. Occasionally, both males and females threaten neighbor groups invading their territory, but they seldom do damage to one another. Group hugs, sometimes when the monkeys are hanging upside down from a branch, are a typical social behavior. These monkeys constantly "neigh," "cluck," and "chuckle" to one another, especially, like the Waltons on the 1970s television show, at bedtime. The muriquis are altogether charming.

That is part of the problem. When Karen Strier began studying them in 1983, she met with a more congenial land-scape than observers of many Old World apes and monkeys. Brazilian primatologists and conservationists quickly became involved in the muriqui project, which was made possible to begin with by the generosity of the landowner and the friend-liness of the local people. Even the local bar owner cooperated

Fig. 1　Muriqui friends in a Brazilian Atlantic Forest fragment.
Photo: Carla B. Possamai.

by allowing the researchers to store fecal samples in his freezer! Still, the disciplinary landscape in which Strier carried out her work was the same as that faced by every other field-worker: there was resistance in the primatology communities of the West to inferences about behavior beyond what could be counted or codified.

Strier's research was conducted according to the scientific protocols of the day, and the 1992 book that culminated her early work, *Faces in the Forest,* is organized according to these protocols. The chapters present, in this order, a physical description of the monkeys; an overview of their shrinking habitat; descriptions of the research station and original methodology; an explanation of how Strier and a team of helpers conducted long-term observations; a discussion of muriqui social behaviors; and finally remarks on the ecology and prognosis for her study species. Strier writes beautiful, lucid prose, and *Faces in the Forest* has rightly been praised and marketed as an example of nature writing for the general public. She takes some slight liberties with grammar: sometimes the monkeys are "which," but other times they are "who." Her word choice occasionally hovers delicately between the literal and the figurative, as the muriquis have "courteous manners" but an occasional "family squabble" breaks out; sometimes the females are "coy," and sometimes vocalizations amount to "discussions."[3] Chapter titles also hint at a literary sensibility at work behind the scientist's mask—for instance, "Early Risers and Other Surprises," "Peaceful Patrilines," and "Life Histories, Unsolved Mysteries." In short, although the study emphasizes empirical data, Strier takes every small liberty she can without breaking the surface of technical, professional style.

However, she has even more to say. Her introductory materials—almost a third as long as the main text in the second edition (1999), from which I quote—reveal that she is absolutely smitten with the muriquis. They have individual stories, some of which are tragic, others comic. They "teach" lessons about themselves.[4] Strier subtly suggests muriqui agency when she writes in the new preface, "The fact that the muriquis have established themselves in . . . comparative scientific literature is a tremendous accomplishment." Six months after her arrival at the study site, she recalls, "the research became more than a dispassionate study motivated solely by scientific questions."[5]

In the introduction, Strier recounts making herself comfortable under the trees where the twenty-three monkeys in the Matão group had settled down for a siesta, only to be interrupted by the loud cries of alarm from a male in the neighboring, but unhabituated, Jaó group. When Nancy, Mona, Didi, and Louise began to respond to his cries, Strier herself became alarmed, but to her surprise, the four females charged the unfamiliar male, neighed and barked, and chased him down the slope until he was well away from their resting place. After they returned to the branches above Strier's head, they

began to embrace one another, chuckling softly as they hung suspended by their tails, wrapping their long arms and legs around each other. Two of the females disengaged themselves from the others. Still suspended by their tails, they hung side by side holding hands and chuckling. Then they extended their arms toward me, in a gesture that, among muriquis, is a way to offer a reassuring hug.

It took all of my scientific training and willpower to resist the temptation—and the clear invitation—to reach back.[6]

The intersubjective potential among the monkeys and between them and the human woman who keeps company with them is perfectly clear—and perfectly represented. All primates instinctively know the meaning of outstretched arms. All primates have arms, hands, and fingers. All primates share a grammar of the body and a lexicon of gesture. With this anecdote, Nancy, Mona, Didi, and Louise—and by extension others in this muriqui society—become three-dimensional, narratively satisfying characters. Strier can be certain of her interpretation because she is certain that primate communication transcends words. By bracketing this information in the introduction rather than trying to synthesize it with the main text, and by emphasizing that she maintained self-control and scientific distance, Strier has it both ways. She remains true to the protocols of current science—and therefore entirely credible to her colleagues in primatology—and she adds information that cannot be integrated into the big picture according to the available protocols.

Barbara Smuts's approach to the problem of characterizing complicated individual animals and accounting for her own feelings is different but equally effective. She splits the personal from the professional, the literary from the scientific. Smuts's book *Sex and Friendship in Baboons* (1985) is a brilliant example of scientific rhetoric, calculated to persuade potential naysayers among her colleagues. Building on the research of Strum and others, Smuts argues that male-female friendship among baboons is an adaptive behavior that benefits females and their offspring by providing protection, males by enhancing opportunities for mating, and everyone by increasing social contacts such as grooming and play. The whole book is bracketed by well-established and noncontroversial factual information about baboons and related species. Within these brackets, almost every chapter is organized as a short monograph, with data at the beginning, followed by summary, discussion, and notes on statistics. The chapters are ordered so that the least quantifiable information (on baboon emotions) and the hot-button issue (male competition) are addressed near the end of the book, after Smuts has already generously described her methodology and established her credibility. Data on interactions among individual baboons appear in the numerous charts and graphs provided in the text (an average of 4.4 per chapter) and fourteen appendixes. Anecdotal information, collected ad libitum and first recorded in field notebooks, is usually set apart in italics and, according to the author, inserted principally to clarify data presented in more formal ways. Nevertheless, these are the most engaging passages for a lay reader and, one suspects, for the scientist as well.

In a new preface to the second edition of *Sex and Friendship*, Smuts claims that her purpose in writing the monograph is to call attention to human responsibility for the planet by recording her experiences with the baboons: "To perceive the planet as populated with billions of such creatures staggers the imagination, but it is true, and if we want the world of the future to retain this richness, we need to become ever more conscious of this reality before it is too late."[7] The carefully crafted literary language of this passage expresses a purpose entirely consistent with science, but it also reveals that Smuts has an agenda beyond the goals of scientific research.

In 1999, the same year that the second edition of her book was published, Smuts contributed to an experimental volume by South African Nobel novelist J. M. Coetzee, entitled *The Lives of Animals*. At the heart of Coetzee's book are two fictional "lectures" about animal rights by a fictional novelist named Elizabeth Costello, framed by a thin plot about Costello's fractious relationship with her family. The volume also contains an introduction by Amy Gutmann, director of Princeton University's Center for Human Values, and "reflections" by the literary critic and "dog lit" specialist Marjorie Garber, animal rights philosopher Peter Singer, religious historian Wendy Doniger, and Barbara Smuts. The contrast between Smuts's earlier publications and her essay in Coetzee's book sheds a little light on the difficulties other primatologists have had in combining the literary with the scientific in autobiographical field narratives.

Smuts's essay develops the theme of her 1999 preface— that is, the emotional richness of relationships among animals. For Smuts, the key to understanding baboons (and humans) is focusing on the *individual within society*. She considers this theme to be underestimated in the fictional lectures of the fictional Elizabeth Costello, who inexplicably fails to mention her own cats. "Under the guise of scientific research" in Kenya, Smuts recalls, she learned that good science consisted of not only collecting good data but also participating in the baboons' way of life—resting when they did and becoming their "humble disciple" as they negotiated the savannah land-scape, avoiding "poisonous snakes, irascible buffalo, aggressive bees, and leg-breaking pig holes." Once she has established the physical setting of life among the baboons, Smuts goes on to describe their social environment, "a system of baboon eti-quette bizarre and subtle enough to stop Emily Post in her tracks."[8] She discovers "what Elizabeth Costello means when she says that to be an animal is to be 'full of being,' full of 'joy.' . . . The default state seemed to be a lighthearted appre-ciation of being a baboon body in baboon-land. Adolescent females concluded formal, grown-up-style greetings with somber adult males with a somersault flourish. Distinguished old ladies, unable to get a male's attention, stood on their heads. . . . Grizzled males approached balls of wrestling infants and tickled them."[9] Finally, Smuts gets down to the business of presenting individual baboon characters (see fig. 2). Some of the characters in the essay show up as data in *Sex and Friendship*, but the intimate and loving detail with which they are portrayed does not.

The most replete of these portraits, of Smuts's "favorite juvenile" as he tenderly examines her hand, reinscribes an often-repeated gesture in primatology literature.[10] Only pri-mates (and gods) have wrists, hands, fingers, and nails: "After touching each nail, and without moving his finger, Damien glanced up at me for a few seconds. Each time our gaze met, I wondered if he, like I, was contemplating the implications of the realization that our fingers and fingernails were so alike."[11] In this essay, Smuts does not hesitate to impute wonder to her baboon "subject" or stumble over the implication that, even without human language, Damien might be able to frame thoughts recognizable in human terms. Damien is more than data: he is a "character" by virtue of his depiction as a sensitive, caring, and thoughtful individual in a freely chosen exchange with another.

Fig. 2   Barbara Smuts with Sultan, a baboon friend.
Photo © Barbara Smuts.

If the literary qualities of Smuts's writing and the graceful interweaving of her own text with Coetzee's did not establish her as a presence in belletristic literature and literary criticism, the story of her relationship with her dog Safi—the "joyful intersubjectivity that transcends species boundaries"—would do so.[12] The term "intersubjectivity" entered language through psychology around the turn of the twentieth century, when it simply meant the capacity of linguistic or other signs to be mutually understood. A sign is "intersubjective" if its meaning is stable and accessible within a social context. The term was more recently retrieved and redefined by feminist psycho-analytic critic Jessica Benjamin. "Intersubjectivity," she writes, "refers to what happens between individuals, and within the individual-within-others, rather than within the individual psyche."[13] Perhaps in hindsight, Smuts recognized that the baboon behaviors analyzed in her book are intersubjective in two ways: the meanings are accessible to her, as well as to other individuals in the gesturing baboon's vicinity, and they also reveal "individuality within others," because, in gesturing, the baboon apparently understands that his or her meaning will be understood by another individual.

It seems quite likely that, all along, Smuts knew she had enough information about baboon behavior to infer complex meanings from what she saw. In participant-observer studies carried out with human subjects by sociologists and anthropologists, "saturation" occurs when researchers see the same phenomena so many times that they are able to accurately predict their occurrence. Smuts certainly reached this point, as most field researchers do, long before her data would be considered statistically significant. In fact, predictable behaviors are part of a search pattern.

Be that as it may, professional distance prevented full intersubjectivity with her research troop. This is not the case, Smuts explains, with Safi the dog, adopted from an animal shelter, "taught" instead of trained, and allowed to become a friend instead of a pet or an object of study. The teaching is mutual, as Safi communicates her own needs and wishes, one of which is to spend as much time as possible outdoors. Smuts is Safi's provider, and Safi has chosen the role of protector on their excursions into the woods. During these walks, Smuts naps whenever she feels sleepy (a detail that, coincidentally, resonates with Karen Strier's experience with the Muriqui female foursome), because she knows that Safi will stay alert. Safi is not a slavishly affectionate animal, but, like the nonhuman primates Smuts has known, Safi offers affection and physical comfort when she perceives that her human companion is distressed.

And thus, with her tales of monkeys and a dog, Smuts completes what is left out of Elizabeth Costello's lecture and her own monograph—that is, the potential of emotional connection across the boundaries of species. At the same time, Smuts completes her own life project of researching baboons according to the protocols of good science and writ-

ing about the experiences, insights, and characters that cannot be recorded in the ethogram or analyzed within the
constraints of the professional publication. Damien and Safi
are three-dimensional, narratively satisfying characters because
Smuts represents them in intersubjective communication with
herself.

Looking at the differences between Smuts's scientific work
and her 1999 essay, it is not surprising that her interests have
shifted toward the study of consciousness and intersubjectivity, which can be analyzed less formally but perhaps more
satisfactorily with dogs than primates. Dogs have coevolved
with humans, and they are far more perceptive and open with
humans than any other species. Anyone who lives with a dog
has a ready-made laboratory on the family couch and does
not have to contend with protocols that prohibit touching.
Smuts has taken this laboratory as seriously as Marjorie
Garber or Donna Haraway, who has also recently shifted
from the study of primatology to the study of canids as cultural configurations.

III

Unlike most of her predecessors in the field of primatology,
Biruté Galdikas published her field autobiography *Reflections
of Eden* (1995) as an unflinching challenge to scientists, conservationists, and casual readers. Since she stands on the
shoulders of Goodall, Fossey, and others who have integrated
the personal and professional in their narratives, Galdikas
can afford to be defiant, even in valorizing intuition and feeling over empiricism. Furthermore, whereas most primatology
literature acknowledges similarities between humans and
other primates, or surreptitiously draws on models or data
from the study of human behavior, Galdikas openly blurs the
boundaries between humans and our primate kin. In doing

so, she completes the literature of primatology, in which the extremes of human and simian being and experience have been censored, suppressed, or hedged about in tactful language. Her book has no index, bibliography, or notes. There are few references to other research that might be relevant to her account of orangutan behavior and ecology; small thanks to the professors who supervised her academic training; and abundant thanks to the members of Earthwatch, whom she lists by name in two pages of very fine print at the end of the volume.[14]

Reflections of Eden is Galdikas's autobiography, constructed around the true life histories of the individual orangutans, free living and formerly captive, with whom she has had the most profound experiences. With the exception of the chapters about her early life in Canada, her sojourn in the United States to finish her dissertation, her place alongside Goodall and Fossey in the world of primatology, and her predictions for the future of the orangutans and the planet, most of this four-hundred-page book narrates the fusion of the author's life with the lives of her study animals. Their lives become hers, and the events of her life outside the forest have meaning only because they frame the story of her relationship with the orangutans. Galdikas provides such detailed and repetitive information about their behavior, biology, and ecology that the reader becomes saturated with facts about these great apes. They are plausible, three-dimensional literary characters with thoughts, intentions, and feelings. Because of the fieldwork already done on other great apes species, the orangutans are believable as empirical representations.

Galdikas's own story, which is difficult to extricate from the ape stories it frames, is raw and deeply personal. Once Galdikas heard about the orangutans and met Louis Leakey at UCLA, every aspect of her life became focused entirely on studying and protecting them. In contrast to Fossey's book, in which all autobiography is eliminated except the bare outline of an early life that ultimately led to the Virungas, or Goodall's early field narratives, which contain only slightly more personal information, Galdikas pulls no punches. She describes falling in love with Rod Brindamour at first sight, her willingness to sacrifice him to her mission of saving the apes, his sacrifices for her career, and his final betrayal of her trust when he fell in love with the young Indonesian woman who cared for their child. Galdikas writes with little restraint about her own sexual nature, her experiences as a mother to both apes and human children, and her second marriage, to Pak Bohap, the much younger traditional Dayak hunter and tracker with whom she now shares her life in a small village near the research station in Kalimantan (southern Borneo). In her single-mindedness, she is less appealing as a narrator than Goodall, less sympathetic than the controversial Fossey, and less guarded than the other scientists whose work I have examined in this book.

Only someone as single-minded as Galdikas could have achieved such a profound understanding of the orangutans, however. By recording interactions among both wild and formerly captive apes for decades—from 1971 to the present—Galdikas has been able to come to an understanding of the typical orangutan's life story. And the longer her research lasts in Tanjung Puting National Park, the more detailed the pattern becomes. Orangutans are the most solitary and individualistic of all primates, including humans. They have startling long red hair, and they are second only to gorillas in size. Their habitat in the swamps and mountain forests of Borneo and Sumatra is even more hostile for researchers than the habitat of chimpanzees and gorillas, and orangutans are more elusive because they spend almost all of their time in the forest canopy. Galdikas's all-absorbing devotion to orangutans familiarized her with two sides of the primate character—individuality and social intelligence. Although orangutan adults do not travel in groups and males live completely alone most of the time, an orangutan infant remains glued to the mother's side for years, objecting strenuously to any attempt on her part to separate. Even adolescent males stay with their mothers far past the age of weaning.

According to received opinion, orangutans are solitary, if not antisocial, beasts and therefore less intelligent than other primates. Galdikas's longitudinal observations, however, allowed her to link orangutan social behavior to their size, caloric needs, and an almost completely arboreal life, which determines the kinds of food they eat—mostly fruit, but also bark, leaves, insects, and, on occasion, other tiny animals. Tropical forests differ from temperate zone forests not only in number of species, which is vastly greater near the equator, but also in their wide dispersal of plants and highly variable plant life cycles. A durian tree in one part of the forest may fruit a month later than another durian tree just a few miles away, for instance. (Durian fruit is large, oval, and creamy sweet; it is beloved by both humans and orangutans.) Since their food sources are spread out over vast tracts of forest, and since each ape has to take in a large quantity of food, it is more efficient for orangutans to travel alone; a group could strip a tree of its fruit in just a few minutes, so living in groups would necessitate more time and energy spent traveling and less on foraging and eating. Both males and females are more food oriented than other apes—males because they are so large (about twice the size of females), females because infants are weaned later than the infants of any other primate species, and both because orangutans eat proportionately more fruit than the other great apes. Unlike chimpanzees, they do not increase their protein consumption by hunting. Unlike gorillas, they do not eat large quantities of leaves.

Galdikas also discovered that, in addition to the mother-infant bond, other social relationships—consortships (sometimes lasting several months) and adolescent friendship groups of three or four individuals (sometimes resulting in lifelong attachments)—are equally part of the orangutan social repertoire. Only the adult males are solitary, because when they meet they compete for females, whose widely spaced birth intervals of about six years mean that females are in short supply (in technical language, females are the "limiting resource"). Males actually avoid meeting one another face to face by means of frequent, dramatic "long calls," which advertise their presence and *their* sexual availability, should any female within earshot be ready to entertain them. The long calls are amplified by the males' huge throat pouches. Pronounced cheek pads, which never stop growing, also distinguish males from females and indicate an individual male's approximate age. Galdikas has speculated that a nonagonistic dominance hierarchy among "cheekpadders" is maintained by the long calls, which are just as individualized as human voices and serve as a kind of signature; a mature cheekpadder's long call gives a young or weak male the option of making himself scarce. In contrast to the males, on the rare occasions when food is plentiful, adult females will sometimes reestablish old friendships for days or weeks at a time, although these old friends are not typically demonstrative.

"Since orangutans do not encounter each other very often," Galdikas notes, "they interact and behave on a scale of time that almost seems to be in slow motion."[15] Proximity alone seems to convey attachment and caring, and embraces by old friends seem to be unnecessary. For the formerly captive apes Galdikas has rehabilitated at Camp Leakey, the orangutan social climate is richer because food is not in short supply— and this richness has helped Galdikas understand that the slow-motion, long-distance relationships of free orangutans demand highly developed social intelligence.

Galdikas learned the most about natural orangutan behavior from Cara and her juvenile son Carl, whose home range near Camp Leakey allowed her to follow them frequently;

once, she followed them daily for a month. Unlike the formerly captive apes, Galdikas felt, Cara was a "peer."[16] She was edgier than most female orangutans and more confident. She never completely stopped hurling branches at Galdikas, not from fear but from irritation at being followed. Once, while following Cara, Galdikas got stuck temporarily between two mammoth logs, whereupon Cara tried mightily to topple a large snag onto the unwanted human voyeur. Had she succeeded, Galdikas would have been killed. Despite her dislike for humans, Cara was popular with female friends and males of all ages among the orangutans—another reason Galdikas was able to learn so much from this particular individual. But Cara had her likes and dislikes among the orangutans, too, and seemed especially to loathe Maud, the adolescent daughter of her friend Martha. Sometimes, Cara was patient with an adolescent male who expressed sexual interest in her and liked to play with Carl; just as often, she chased him away.

Galdikas learned about orangutan consortships from watching Cara. She followed Cara closely enough during her pregnancy with Cindy to collect solid data on orangutan gestation, birth, and early infant development stages. Because she was pregnant, Cara weaned Carl, who appeared to be only four years old, much younger than the normal age of six or seven. Since young orangutans strenuously object to weaning, it can be a vicious process, but sometimes Cara seemed to take pains to reassure Carl that he was still under her protection.

Galdikas's observations of Cara also taught her how orangutans use the forest. She watched Cara crack and eat huge banitan nuts, which cannot be opened even with an ordinary hammer. Cara broke open termite mounds and devoured the contents, and she made umbrellas from large leaves. Galdikas watched her construct comfortable sleeping nests and, on more than one occasion, make dummy nests to throw unwelcome companions off track. However, the most important, and heart-wrenching, information Cara provided was what happens when the forest and its inhabitants are under pressure from erratic weather patterns and diminished boundaries. Over a period of several months, Galdikas observed a decline in the orangutan's appearance, which she at first attributed to the birth of the infant Cindy. Then Carl became emaciated quite suddenly and died. Galdikas had not yet solved the mystery of Carl's illness when she discovered Cara holding Cindy's tiny body, literally mummified because Cara's constant grooming had kept away the flies. Finally, well after Cara herself succumbed to the malady that took Carl and Cindy, Galdikas began to understand what caused the family's illness and death: it was starvation. A drought year, during which there was little fruit, had been followed by a year of unusually heavy rains, which rotted the fruit. The orangutans could not survive on the bark and insects that they sometimes ate as dietary supplements. The timing could not have been worse for Carl, who had just been weaned a little too young and was learning by degrees to find his own food, or for Cara, whose caloric needs were higher than usual because Cindy had not yet learned to consume anything other than her mother's milk. Cindy starved, of course, because there was too little milk. These apes died not so much as the result of "the unseen hand of natural selection," but because of "the visible, iron-fisted hand of human progress. . . . Perhaps the destruction of a forest fifty or a hundred miles away from Tanjung Puting played a role in Cara's death . . . by increasing the orangutan population around the reserve and hence competition for food. . . . Perhaps [the] unusual weather pattern was related to global warming, and Cara was a victim of human interference in the biosphere."[17] In any case, for Galdikas, nothing in all her life "devastated me more than Cara's, Carl's, and Cindy's death."[18]

As Galdikas's account of Cara's family suggests, the appeal of *Reflections of Eden* does not depend entirely on detailed descriptions of the apes themselves. And the challenge mounted by the book is more than an emotional plea or a dramatic reinterpretation of what is known about orangutan existence. Despite Galdikas's careful explanation of the adaptive bases of orangutan behavior and passionate discussion of the logging that is quickly decimating their habitat, the picture of the apes that emerges in this book is as raw as her own autobiographical confession.

It is clear from the very first page that these apes are sometimes hard to love, even for Galdikas. Akmad, a former captive rehabilitated and released from Camp Leakey, returns with an infant after a long absence. They sit near Galdikas at the edge of the forest. After a few minutes of mutual silence, Galdikas asks Mr. Achyar, the camp feeder, to stop preparing pineapple for a moment and help her arrange the infant for a photograph. Although Akmad knows Mr. Achyar well— orangutans' memories of individuals are evidently at least as accurate as human memories—"Akmad recoiled. Baring her teeth, she exploded. Her hair went erect, tripling her size, and she lunged at Mr. Achyar, her fangs glistening. . . . Leveraged, taut muscles provide even female orangutans with the strength of perhaps five men. An orangutan female's teeth can rip off a person's scalp or arm. Had Mr. Achyar not been so agile, he would have been badly mangled. . . . Her point made, she simply sat down and picked up the pineapple she had been eating. Her face was once again expressionless."[19] Galdikas herself does not hesitate to continue arranging the infant for a photograph. The strongest bond among orangutans is between mothers and daughters—and she is, in a literal sense, Akmad's mother.

Quite early during her research in Borneo, Galdikas and Brindamour started urging local authorities to confiscate the young orangutans many well-off people kept illegally as pets. Once taken, the apes ended up at Camp Leakey. An infant male they named Sugito was the first of these youngsters. Galdikas's experience with the formerly captive infants, juveniles, and adolescents is interesting in terms of feminist theories about gender roles, particularly the role of the mother, and primatology research on parenting. In *Mothers and Others*, published more recently than Galdikas's book but relevant to her depiction of motherhood, Hrdy argues that human parenting is by nature similar to the allomothering patterns she observes in bonobos. Unfortunately, she concludes, contemporary Western cultural expectations isolate human women with their infants, to the detriment of both mothers and children. (Hrdy's analysis is remarkably similar, and seems to be indebted to, Adrienne Rich's analysis of motherhood as a cultural institution in *Of Woman Born* [1977].) The isolation Hrdy describes parallels the isolation that, according to Galdikas's observations, is completely natural for orangutan mothers and children. If human mothers are inadequate to the demands of twentieth-century motherhood because of social expectations, Galdikas finds that she is inadequate for the role of orangutan motherhood because of species requirements. Although she has been prepared by twentieth-century American society to be the sole caregiver needed by the baby orangutans, Galdikas has, on the other hand, been fitted by her evolutionary history to hand off babies to other caregivers, as the bonobos do.

But the little orangutans will not allow themselves to be passed around, and consequently their "mother" is often a wreck: "At times, I would almost forget he was there as I went about my daily routines. Other times, however, an uncontrollable rage would well up within me. I would pull Sugito off my body and run away. He would screech piercingly and run after me.... When I persisted and did not let him back on my body, he threw a temper tantrum. . . . Great gushes of guilt would flood over me at the sight and sound of Sugito's heartwrenching displays."[20] Knowing that true orangutan motherhood takes place within a context of different stresses does not keep Galdikas from enmeshment with this infant, who bites her husband and masturbates in his ear, urinates and defecates in the couple's bed every night, refuses to let her bathe or even change clothes, destroys everything within reach, and insists on sucking her thumb until it is raw and bleeding. Needless to say, Sugito is jealous of the other infants who arrive at Camp Leakey during the months and years that follow. In spite of his behavior, Galdikas writes, "I grew to treasure the little orange creature whose eyes fastened on mine first thing in the morning and who attended my every move. Sugito gave me the highest compliment any man or woman can receive. To him, I was, by far, the most important individual in the universe."[21] Sometimes, she literally forgets that he is not human and she is not orangutan. This mother-infant pair bond would have delighted Freud, since it confirms, many times over, now-suspect Freudian theories about human sexuality—or would, if both Sugito *and* his "mother" were human.

Later, after she has been following free-living orangutan mothers and offspring for many years, Galdikas comes to understand normal orangutan infancy more deeply and is able to change her rehabilitation methods slightly by encouraging the older orangutan children to form affectional bonds with younger ones. But she never changes her account of the general structure of the mothering role, first established by her relationship with Sugito. Indeed, Brindamour's many photographs of his wife with the infants who came to Camp Leakey during the following years—playing with them, holding them close, struggling to walk with several of them dangling from her body at the same time—appear in *Reflections of Eden*, as well as the two articles she wrote for *National Geographic*. Goodall and Fossey have experienced something like a maternal instinct in relationships with their study animals, and Fossey took on this role briefly with the famous Coco and Pucker before she was forced to part with them. Galdikas revels in it. This insistence on motherhood consigns her to a position outside the boundaries of the science of primatology and, insofar as the role potentially interferes with normal behavioral patterns for the animals, even challenges the parameters of conventional conservation practices.

If orangutan behavior is unattractive in these anecdotes of Sugito's infancy, it is even less appealing elsewhere in Galdikas's story. She notes in her book that one of the apes murdered another, and in the 1980 *National Geographic* article "Living with the Great Orange Apes," she reveals that Sugito was the murderer. She uses the word deliberately, to indicate intentional wrongdoing. A few years after Sugito is "weaned," Galdikas explains, a small juvenile female is found dead by the other juveniles. Alerted by their cries, Galdikas arrives on the scene and notices that Sugito is the only one who will not look at her or the body. She reads his body language as guilt. (In spite of his crimes and misdemeanors, Sugito remains so precious to Galdikas that he is allowed to stay in the camp, until Brindamour decides to remove him without consulting her.)

Whereas most of Sugito's aggression is directed at the other apes, other formerly captive orangutans can and do threaten the humans who care for them, as Akmad's story shows. More serious than Akmad's behavior—because the aggression is not merely threatened—is the behavior of Gundul, a male orangutan whose owners kept him as a pet much longer than is normally feasible. A normal pattern among adolescent male orangutans, Galdikas discovered, is forced copulation with the adolescent females associating with them in temporary social groups. Galdikas calls this typical behavior "date rape" and observes that adult male orangutans do not engage in it. By the time he is released from captivity, Gundul's sexual aggression has become directed at humans rather than other apes, and he sexually attacks the wife of one of the staff members at the camp—face to face, in the typical orangutan position for sex. Galdikas explains that orangutan "rape" is not wrong or aberrant, because it is part of the normal behavioral repertoire and because orangutan female victims may feel annoyed but not violated or guilty. Still, these stories do not show orangutans in an attractive light. Since Brindamour is the dominant male in the camp and Galdikas the orangutan mother, they are safe. But in addition to their sexual transgressions, the former captives often chase, threaten, and bite students, visiting forestry officials, and even members of the support staff. It turns out that Poe's story of the marauding orangutan in the Rue Morgue is not far off the mark.

How, then, does Galdikas make a case for the protection and preservation of habitat for a creature that only a mother could love? Naturally, readers who pick up *Reflections of Eden* are already science- and conservation-minded. Galdikas argues that orangutan habitat—the vast, formerly impenetrable forest of Sumatra and Borneo—is disappearing at an alarming rate due to logging and climate change, and, as all environmentalists know, the loss of tropical forest further endangers the health of the entire planet. Galdikas's descriptions of the orangutan life cycle within these threatened tropical forests not only carry considerable weight but provide material data for her composition of new arguments for readers interested in protecting nature.

Still, her orangutan characters inspire sympathy only within the context of another kind of argument—a religious argument that is developed indirectly, through the deep structure of the book and a web of figurative language. Confessional autobiography as a literary genre developed within early modern Christian culture. Written in the fourth century, the most famous of these books, Saint Augustine's *Confessions*, established a pattern that would be followed many times. Augustine's insights about spiritual life are supported by his account of a dissolute early life before he gave due respect to his mother's religion, as well as the changes experienced after he set himself upon her path. Augustine's autobiography has been enormously influential, especially in the development of Catholic doctrine about sexual sin and the development of autobiography in Western literary tradition.

Women, too, wrote spiritual autobiographies. *The Book of Margery Kempe*, authored by a late fourteenth- to early fifteenth-century mystic who reached an agreement with her husband to live together in celibacy because of her growing conviction that sexuality was a sin, is one of the best known. When the neighbors started to discuss the terms of their marital arrangement, Kempe became a solitary itinerant preacher, performing corporal works of mercy and trying to make converts to her eccentric rule wherever she went. Kempe's manuscript was lost shortly after her death and not rediscovered until 1934, but in the seventeenth and eighteenth centuries, as the Reformation branched off into many other denominations and evangelical movements, Kempe's work— both her writing and the example of her life—was paralleled by scores of religious leaders, both women and men.[22]

The confessional form eventually branched off into secular autobiographies, such as the post-Christian *Confessions* of Jean Jacques Rousseau; nineteenth-century narratives of slavery and Indian captivity; twentieth-century texts including Simone de Beauvoir's *Memoirs of a Dutiful Daughter* and Maya Angelou's *I Know Why the Caged Bird Sings*; and numerous stories of addiction and recovery written in the late twentieth century. De Beauvoir, Angelou, and a host of other twentieth-century writers recall moments of pain, victimization, and guilt, which they offer as lessons from which others can learn. The genre has remained alive and well through the turn of the twenty-first century, too, in books such as Marya Hornbacher's *Wasted*, a wrenching account of disordered eating, and Naomi Wolf's *Promiscuities*, a feminist analysis of youth culture that is based, in part, on her own less than stellar behavior as a high school student. All of these accounts have, if not an overtly religious cast, a spiritual dimension.

Galdikas's work partakes in the religious and confessional aspects of this autobiographical tradition: the experience of sin, for her, is a gateway to understanding. Galdikas's "sins" are strangely reminiscent of Margery Kempe's. For the sake of goals she deemed more important than her relationship with Brindamour, she failed to nurture her marriage and almost deliberately sacrificed him for the apes. For example, when Brindamour was diagnosed with a serious lung infection during a visit to the United States, she insisted that he return to Indonesia with her and get treatment in Jakarta. Knowing that there was an attraction between her husband and Yuni, a young Balinese woman the couple knew in Jakarta, Galdikas hired Yuni to care for their son, Binti, anyway; she wanted more freedom for both herself and her husband to work with the orangutans and was willing to risk her marriage to get it. Galdikas overlooked every sign and signal as Brindamour neared the breaking point. She confesses to social gaffes in the United States and Indonesia and bad judgments even in her work with the orangutans. She chose not to treat or provision Cara and her family, and they died. Although many field-workers would agree with her decision, she regrets it. Galdikas even admits that some of her successes have been the result of ignorance rather than planning.

According to the Christian model, understanding ideally leads to devotion to a cause. In Augustine's case, the cause was theology, and he sacrificed all his secular privileges to pursue it. Kempe sacrificed material and marital security for her religious convictions. Even for de Beauvoir and Angelou, Wolf and Hornbacher, pain and guilt have enhanced their abilities to plumb greater depths as writers and thinkers. Whatever the cause, it need not be equally appealing to everyone, but it is assumed that every reader can understand the devotion to a cause. Galdikas's cause, of course, is saving the orange apes of Indonesia and with them, she hopes, the planet.

Although Galdikas divulges her Catholic childhood and her parents' faith, her own metaphor for God or Providence is the cat's cradle that Louis Leakey expertly weaves again and again when she visits with him at Jane Goodall's apartment in London. A nervous tic, a child's game, an entertainment? For Galdikas, the pattern Leakey created with a simple length of string came to represent the web of cosmic connections that led her to Borneo and kept her there. Like the Christian saints who minister to the poor, the ill, or the criminal elements in society, Galdikas has her own mission to the orangutans of Borneo, and she is willing to sacrifice everything for them. According to Christian logic, the son of God was sacrificed for us not because we deserve it, but because we do not. God loves us not because we are loveable but because God himself is love. The orangutans have redeeming qualities, but within Galdikas's Christian framework, their very recalcitrance becomes a reason to love them.

For Galdikas, the orangutans are not just figuratively but literally representative of a prelapsarian primate past, and this argument for their protection, too, is a religious one. They are "reflections of Eden." Since they are not human, they do not subscribe to or understand human morality. Deviant orangutan behavior, such as Gundul's sexual fixation on human women, arises when the animals become accustomed to human ways. Sugito's hell-raising destruction of foodstuffs, supplies, clothing, and even buildings comes about simply because orangutans do not have or understand material culture. Thus, their misbehavior becomes the result of one more human crime. Sugito's murder of the young female and his apparent guilt suggest that orangutans do, in fact, have a rudimentary moral sense like ours, but if human interference had not created Sugito's unnatural social life, he would not have killed. Whatever they do, the orangutans are blameless:

Our appearance on the earth was relatively recent; orangutans are far older, as a species, than we are. I wonder, when Homo erectus *strode into Asia: were orangutans watching from the trees? It is a humbling thought.*

Our departure from Eden allows us reflection—reflection on our origins and our relations to other creatures, reflection on good and evil, and, ultimately, reflection on the possibility that we are engineering our own extinction. Never having left Eden, our innocent pongid kin are not burdened with this knowledge and the responsibility it entails. Looking into the calm, unblinking eyes of an orangutan we see, as through a series of mirrors, not only the image of our own creation but also a reflection of our own souls and an Eden that once was ours. And on occasion . . . with an intensity that is shocking in its profoundness, we recognize that there is no separation between ourselves and nature. We are allowed to see the eyes of God.[23]

As Goodall's late work likewise shows, if one extreme in the literature of primatology is Harry Harlowe's infamous slew of laboratory experiments depriving rhesus babies of maternal contact, the other extreme is the spiritual contemplation of what nonhuman primates mean to us and what they *are* as a critically endangered ontological presence in the world.[24]

During the late 1970s and early 1980s, some of the rehabilitated apes at Camp Leakey, including the poster child Princess, were taught sign language by graduate students Gary Shapiro and Benny Ismunadji. Galdikas viewed these experiments with interest, and in her 1980 *National Geographic* article, she commented on the possibility of learning more about the orangutans' perceptions and feelings by communicating with them through American sign language. However, by the time she wrote *Reflections of Eden* more than a decade later, she apparently did not feel that she needed the information this could provide about the orangutans' subjective experiences.

In Galdikas's more considered view, even though humans may sometimes overlook the species boundary, we are irrevocably separated from orangutans by bipedalism, symbolic language, and sin. The human toddler Binti learned sign language more quickly than the apes, who were older but less interested and adept. Instead of signs and symbols, Galdikas herself relies on the more embodied Edenic language of gesture, as well as her total immersion in the Bornean jungle. This approach helps her understand the orangutans' experiences intersubjectively, from the inside out. From the forest and the apes themselves, she derives absolute confidence in the accuracy of her inferences and an absolute conviction about the rightness of her cause. She makes the claim over and over again: Borneo is Eden, the apes recall our own Edenic past, the apes themselves are Eden. When Leakey called her an angel, along with Goodall and Fossey, Galdikas took him literally.

Reflections of Eden does not respond to the essential problems of epistemology and rhetoric as most other primatologists respond to them. The book is Galdikas's unique way of asserting a truth and preaching a gospel, and it is one of the most dramatic statements of the twentieth century about the importance of primatology as a planetary project for the twenty-first. Since 1995, Galdikas has written and edited numerous other books for the general reader, or for an audience, including scientists, interested in an overview of contemporary issues rather than the close reading of data. She continues to work toward "saving Eden," as she entitled the last chapter of *Orangutan Odyssey* (1999). She fights for the preservation of the orangutans by rehabilitating former captives, supporting the work of primatology students at Tanjung Puting National Park, fostering ecotourism, and participating in international legal efforts to preserve the apes and their habitat. Her extremism, although different from Fossey's, has made her controversial.

But these are desperate times.

CHAPTER SIX

Primatology and the Carnival World

Invest me in my motley; give me leave
To speak my mind, and I will through and through
Cleanse the foul body of the infected world,
If they will patiently receive my medicine.

—SHAKESPEARE, *As You Like It*

I

Literary genres are worldviews. In adopting a particular story type, an author reveals her or his deepest intentions, beliefs, values, and assumptions. Romance valorizes individual identity and experience. Comedies emphasize beginnings, tragedies call attention to endings, and soap operas dwell on the impossibility of establishing clear beginnings or endings in everyday life. *A Primate's Memoir: A Neuroscientist's Unconventional Life Among the Baboons* by Robert Sapolsky can be read as a burlesque or parodic bildungsroman. The bildungsroman, a story of personal development thoroughly situated within a social context, combines the motivations of the romance with those of the realist novel.

A parodic bildungsroman is only possible at certain moments in history—in this case, within the history of primatology. Most of the narratives considered so far in this study feature an important element of the bildungsroman: the narrative arc, which, to some degree, illuminates personal and professional development. Between 1960 and 1990, the field burst open alongside postcolonial development and habitat shrinkage, the barriers for women in science started to tilt away, and the science of primatology itself changed drastically with new data, new methods from field studies, and ethical pressures from within and outside the discipline. The bildungsroman no longer seems quite relevant or believable.

The work under consideration here resembles Shirley Strum's primatology "novel" in subject matter and setting, and Sarah Blaffer Hrdy's in its indeterminacy, but in the decades between their experiences and Sapolsky's, the field changed drastically. The landscape is more crowded, the humans who share it with the baboons more corrupt, and the ethical questions more vexing. Like Biruté Galdikas, Sapolsky is constrained as a scientist and storyteller by the damaged world in which he finds himself and by his own imperfections, so the confessional elements of his story rise to the surface. But unlike Galdikas's book, his narrative overtly situates itself within both primatological and literary traditions.

Moreover, like other writers of primatology narratives, Sapolsky makes a point of establishing his credentials as a reader—and his favorite authors are a cue for reading what he writes. Thomas Mann, whose probing psychological novels exemplify the twentieth-century bildungsroman, is one of his standbys. Sapolsky claims that the "worst literary mistake" of his life was taking along Mann's *Joseph and His Brothers*, with its endless descriptions of desert scenery, for light reading when he hitchhiked through the Sudan, where endless desert scenery in fact turned out to be the least of his problems.[1] Less humorously, Sapolsky alludes to Mann's *The Magic Moun-*

tain, the story of the emotional development of a young man trying to recover from tuberculosis in the microcosm of a fashionable sanatorium. (The reference to this story about the etiology and treatment of tuberculosis, it turns out, is no more coincidental than the choice of literary genre.)

"I don't basically think of myself as a writer," Sapolsky commented in an interview about his writing. But he goes on to say that, for him, writing is sometimes easier than science, because it allows for communication when other means of connection have been impossible.[2] Writing for the nonspecialist reader also allows for communicating the intangibles that do not appear in the professional scientific literature. An entirely different protocol is in effect, and, like any other conscious belletristic writer, Sapolsky places every element in the story deliberately. When revising *A Primate's Memoir* for publication, he set up a flowchart of episodes in order to establish the interlocking chronologies of his own story and the history of his study troop. In an effort to create a coherent narrative, he says, "I went through this very mechanical algorithm of listing all the baboons [and] where they were introduced."[3] This approach to structure in fact resembles Charles Dickens's method for planning out his complex triple-decker plots for serial publication. Regardless of whether "writer" is the first element of Sapolsky's self-definition, a writer he certainly is.

Sapolsky's memoir is divided into four parts, the titles of which allude to the developmental stages of a baboon, or a human: "The Adolescent Years: When I First Joined the Troop," "The Subadult Years," "Tenuous Adulthood," and "Adulthood." Sapolsky's field research in Kenya, alternating with periods of research in the laboratory back home, is marked according to the "reigns" of various alpha males—notably Solomon, Uriah, Saul, and Nathanial. The stories of other individuals, male and female, are told against the backdrop of their struggles for dominance. The reigns of the alpha males proceed according to a fairly predictable pattern until a human-made crisis drastically changes life in the troop, perhaps permanently. Just as Sapolsky's experience makes him a sadder and wiser man, the baboon troop, as a result of human contact, is eventually forced out of "innocent" lifeways, which have evolved over thousands of years, into unexpected patterns.

Sapolsky does not see his own story as an isolated case. He gradually builds the argument that his particular experiences represent the pratfalls, joys, griefs, dangers, and ethical tensions in the lives of many scientists, and that the life of a contemporary scientist is hedged about with extradisciplinary circumstances impinging on and sometimes compromising his work. Thus, behind and alongside the individual experience of the author are both a story of the baboons and a story about the contemporary international scientific community. All of this takes place against the backdrop of political, economic, and cultural turmoil in present-day Africa.

Primatology is a multidisciplinary field, and part of Sapolsky's story concerns the tension of his position among the various scientific disciplines—most obviously between experimental science and baboon field research, but also between biochemistry and ethology, two disciplines that seem, and often are, in conflict because the first is invasive and the second ordinarily conducted at a respectful distance from the animals. Sapolsky's work on the connections among stress, stress hormones, behavior, and social status underscores the fact that the boundaries between laboratory science and field science are useful but artificial, dotted lines instead of solid. The choice of the bildungsroman structure for *A Primate's Memoir* affords a way to understand and describe the situation of his particular work across these disciplinary boundaries.

But the book is also a parody. Sapolsky's view of himself is relentlessly and hilariously zoomorphic, thus standing the traditional scientific admonition to avoid anthropomorphism on its head. "I joined the baboon troop during my twenty-first year," he begins. "I had never planned to become a savannah baboon; instead, I had always assumed I would become a mountain gorilla."[4] This statement is not just a clever hook, although it serves that purpose: it is the central trope in a narrative loaded with irony and paradox throughout. And even though the jokes surely helped make the book an entertaining best seller, Sapolsky's relentless parody also serves a serious purpose as part of a critique of his discipline and the larger world in which primatology is practiced. In *Thinking Animals: Why Animal Studies Now?*, Kari Weil calls this approach "critical anthropomorphism." Sapolsky's generic choices may also reflect uncertainty, not about the validity of his observations and conclusions but about his place in the history, demographics, and practices of primatology.

II

In his critical classic *The Visionary Company* (1971), Harold Bloom has argued that poets and other artists are always anxious about their cultural patrimony—that they must somehow learn from their forefathers and yet rebel enough to distinguish themselves. In this way, suggests Bloom, poets participate in what Freud famously called a "family romance," in which the son emulates the father, rebels against him, and has to figure out a way not to kill him off directly if he wants to consider himself civilized. (In Freud's version, the son also wants to own the mother—more on that later.) Such tensions are not unlike the stresses that Sapolsky discerns in baboon dominance struggles and, indeed, suggests with reference to his own life. If the "sons" are writers, claims Bloom, they manage to distinguish themselves against their literary forefathers by "misprision" and more or less subtle plagiarisms. Misreading and theft leave the son with "anxiety of influence," a sense of belatedness and nervousness about his own identity as a poet or—although Bloom doesn't say so—a painter, composer, or scientist. Like many literary and scientific metaphors, the Bloom/Freud family romance (which is not the same as the quest romance) has enough truth in it to make it useful.

Feminist literary critics Sandra Gilbert and Susan Gubar do not argue with Bloom's theory about literary men, but they ask, "What about the women?" The family romance requires not the daughter's rebellion against the father but her accommodation to him. So, according to the theory of Gilbert and Gubar, women artists and intellectuals experience internal division, sometimes to the point of madness. Their influential book, *The Madwoman in the Attic* (1979), is about creative madness, and sometimes the literary semblance of madness. This pattern may sound familiar in the annals of twentieth-century primatology.

In *Honey-Mad Women* (1988), Patricia Yaeger has argued that this madness and fakery are not always sad or bad. Women writers sometimes take old literary forms and turn them upside down and inside out, in what Bakhtin calls the "carnival mode." For this element of his literary theory, Bakhtin looks back to the classical and early modern periods of Western cultural history, when carnival meant throwing out the rule book; crowning peasants king for a day; eating, drinking, and fornicating to excess; and celebrating life and license. In carnival, the boundaries of the body are understood as permeable, and those bodily functions normally kept under control are released and celebrated. Most important in literary terms is the permission conferred by carnival to say things that would be disrespectful, treasonous, or blasphemous in any other situation. In carnival, Aristotle wears a dunce cap. Bakhtin sees traces of the carnival world in all kinds of cultural forms, especially in satire, parody, and anti-romance. Carnival is not simply an entertainment: it provides release for emotions and information that are normally bracketed or hidden.[5]

Sapolsky is a serious neuroscientist whose work has appeared in academic journals, respected trade publications such as *Scientific American* and *Natural History*, and several other books for the general reading public. Before writing his memoir, Sapolsky had already reached a wide audience with his articles on health, which could only have been written by someone with a sympathetic and specialized understanding of human behavior, biochemistry, and the stresses of industrialized life. *Why Zebras Don't Get Ulcers*, a study of contemporary health issues explained in terms of primate stress reactions and stress hormones, is not a funny book, although Sapolsky's self-deprecatory humor flashes out at times. Despite arguments replete with technical information that is sometimes challenging for the lay reader, the book has gone into its third edition. Sapolsky's lecture presence is similar to the ethos of the book: he is warm, informative, and humorous without being the last comic standing, and he is evidently committed to convincing his audiences, usually college students, that they should reflect more carefully about their health and lifestyles.

So why, in *A Primate's Memoir*, does he present himself as a buffoon? Why are the baboons characterized as smaller versions of himself—a self-described atheist from an Orthodox Jewish family—and named for the Old Testament patriarchs and matriarchs? Why does his memoir participate in the carnival world? Why does the story veer sharply, about two-thirds through, from comedy to sadness and high seriousness? These are questions about literary genre, but the answers to them reflect upon the history and practices of primatology.

Sapolsky's place in the family romance of primatology is significant. Goodall knew perfectly well that at first few primatologists cared what she was doing, and she realized a little later that her standpoint was, in fact, oppositional. If her writing does not descend into the madness postulated by Gilbert and Gubar or the crazy comedy described by Yaeger, she still writes eloquently of the emotional trauma resulting from the hostile reception of her most important discoveries. At that point, in the 1960s and 1970s, primatology was as indisputably male dominated as any other science, but Goodall's status as an exceptional woman may have helped her career more than hurt it, and her oppositional position actually enabled her to change the field itself. She was thus a role model and big sister for Fossey, whose research into the lives of the virtually unknown mountain gorillas, beginning less than a decade after the start of Goodall's chimpanzee study, confirmed many of her own insights about chimpanzees. There seems to have been little, if any, rivalry between these two, or between Goodall and Galdikas, Leakey's third protégé, though Galdikas did regard both Goodall and Fossey as formidable models.

Primatology has been perceived by many in the field as a female-dominated discipline since the 1980s. Many women entering the field seem to have experienced not anxiety of influence with regard to their male predecessors, but an empowering female affiliation, and less frequently rivalry, with their female predecessors. Women in the field have also been media darlings. The proposition of female dominance is arguable, however. Linda Fedigan, an influential editor and researcher specializing in capuchins and snow monkeys, and Shirley Strum, whose scholarship includes publications on the history and philosophy of science, together argue against the notion of female dominance in *Primate Encounters*.

In any case, at the start of his career, Sapolsky found himself in a rare and disconcerting position for a man of science. He had to build his own scholarly project on theories, methodologies, and a body of knowledge initiated and shaped by men, and then challenged by a number of women who disagreed with them on important points of theory, methodology, and scholarly protocol. Sapolsky could be considered a man in a woman's world in a man's world.

The disagreements within primatology partly reflect the kind of generational shift that takes place in any field of knowledge, but the fact that the shift has been led by a relatively large number of women, newly present in primatology, contributes to the perception that the field is female dominated. The ostensible topic of the following passage from "Monkeyluv," the title piece of Sapolsky's 2005 essay collection, is the generational shift in thinking about dominance in male primates, but the asides suggest a sense of marginality:

The revolution was the discovery of "female choice," the wildly radical notion that females had some say in the matter. Maybe this had something to do with there having been a transition, such that the best primatologists around were female, and with their looking at the behavior of their animals without that linear-access bias [the assumption that alpha males always get their way with the females]. What was obvious was that some females didn't just passively wind up mating with whichever hunk strutted forward. Being half the size of males in many of these species, females couldn't convince a male they didn't favor to get lost by beating on him.

But they sure could fail to cooperate.[6]

Sapolsky goes on to say that female baboons often pick the "nice guys" for copulation and cleverly manipulate all the males to get their way. (He includes several anecdotes about female choice in *A Primate's Memoir*, without framing the subject as part of a generational shift in the discipline of primatology.) Clearly, Sapolsky is one of the nice guys—most of the time—giving credit where it is due.

Sapolsky always knew what he wanted to do: when he was a child, he wrote fan letters to Irven DeVore and Stuart and Jeanne Altmann, all of whom conducted early baboon studies in Africa, but by the time he got to graduate school, the practice of primatology had changed. So, if a man is used to having to assert himself against the patriarchs, as Freud and Bloom say men must, how does he distinguish himself against the patriarchs and the oppositional matriarchs simultaneously? It would be difficult to rebel against those whose work you emulate and from whom you learn your craft (Jeanne Altmann and Shirley Strum have, in fact, opened their field sites to Sapolsky) without risking your credibility or being churlish. It would be harder still to rebel against the rebels. There is only one Jane Goodall, one zero in the Tarot deck, one innocent fool who sets out on the journey of self-becoming without antecedents and without fear. There is only one Dian Fossey, who lived with the gorillas Sapolsky himself dreamed of studying. Starting over is not an option. In this case, the Freudian paradigm, which includes the son's desire to own the mother, is also not an option because no one "owns" these matriarchs in the field, so ownership can't be passed along. At most, given Sapolsky's position within the family romance of primatology, these women are big sisters.

Being a nice guy helps. But if you're a man of science in a world perceived as woman centered, your own romance becomes an anti-romance. The plot may be similar, but instead of being a hero, you sometimes present yourself as a joker. You have helpers along the way, but they are often figures of fun instead of reverence. The external and internal obstacles are real, but they are also ludicrous. The sacred is profane. The rules are broken. The body is permeable. Human becomes animal and vice versa. Sapolsky gets his laughs by misreading, misprision, parody, and self-parody.

III

In *A Primate's Memoir*, as well as Sapolsky's other pieces for the lay reader, men are frequently comic figures, and Sapolsky himself is the most comic of all. He is not a heroic scientist errant but a fool in the Shakespearean sense, in order, finally, to tell truths most unpleasant to hear. "Motley's the only wear," says Jaques, Shakespeare's cynical exiled courtier, who envies the clown suit and, even more, the carnival license possessed by the bawdy fool Touchstone. In his memoir, Sapolsky's resemblance to both Jaques and Touchstone is more than superficial.

One of the most visible layers of the buffoonery is Sapolsky's identification with the baboons he has come to study (see fig. 3). Benjamin, a male transfer into Sapolsky's study troop, is a simian analogue of the author himself:

Still just emerging from my own festering adolescent insecurities, I had a difficult time not identifying utterly with Benjamin and his foibles. His hair was berserko. Unkempt, shocks of it sticking out all over his head, weird clumps on his shoulders instead of a manly cape that is supposed to intimidate your rivals. He stumbled over his feet a lot, always sat on the stinging ants. . . . Every time he yawned . . . he had to adjust his mouth manually. . . . He didn't have a chance with the females, and if anyone on earth had lost a fight and was in a bad mood, Benjamin would invariably be the one stumbling onto the scene at the worst possible moment.[7]

One day, several years after Sapolsky's first meeting with him, Benjamin actually solicits Sapolsky's help when he is about to be the victim of displaced aggression in a dominance dispute. "In the name of all my professional training and objectivity, but to my perpetual shame," he writes, "I had to pretend that I didn't speak the language and had no idea what he was talking

about."[8] A common conflict, apparently, among field primatologists. Needless to say, the clueless Benjamin becomes his
favorite baboon, and this passage about Benjamin typifies
Sapolsky's attitude about all the baboons: they *are* like us. In
fact, the resemblance between humans and simians is what
fuels primatology, despite the field protocols and professional
style that muffle both the resemblances and the emotions
inspired by our fellow primates—and which Sapolsky flouts
through the style of this passage, though he tries not to do so
in his academic science writing.

Pointing out the similarities between himself and his
baboons in such graphic and personal terms is part of the
rebellion Sapolsky mounts through the burlesque. The protocols operate against the appearance of anthropomorphism, so
Sapolsky responds with zoomorphism, as in his identification
with Benjamin. At the end of the first chapter in *A Primate's
Memoir*, Sapolsky mounts a full frontal challenge to scientific
language: "Debates rage among animal behaviorists as to the
appropriateness of using emotionally laden human terms to
describe animal behaviors. Debates as to whether ants really
have 'castes' and make 'slaves,' whether chimps carry out
'wars.'"[9] In spite of his own reservations about using anthropomorphic language, Sapolsky ends the chapter with an incident of baboon "rape." Taking out his aggression on someone
who cannot defend herself, the just-deposed alpha male Solomon attacks Devorah by an acacia tree. In using the word
"rape," says Sapolsky, "I mean that she had not presented to
him, was not behaviorally receptive or physiologically fertile .
. . that she ran like hell, tried to fight him off, and screamed in
pain when he entered her. And bled. So ended the reign of
Solomon."[10]

Sapolsky's language in this passage represents a swipe
against the patriarchs of the discipline, who resist language
that reveals the deepest similarities between humans and
other animals. And although it is not my intention to give a

Fig. 3 Robert Sapolsky with a baboon friend.
Photo: Lisa Share-Sapolsky.

full-blown reading of the Freudian family romance as it plays
out in his book, it is hard not to read Sapolsky's choice of
names—Saul, the first king of Israel; Solomon, the wise ruler
at the height of Israel's historical power and prominence;
Bathsheba, his bewitching lover; Devorah, the great military
leader—as another kind of swipe against the conservative
patriarchs of his own family. In any case, the comic irony of
baboons bearing Old Testament names contributes to the
general hilarity of Sapolsky's narrative. Accounts of their

battles for ascendancy can be read not only as realistic descriptions of baboon social behavior but as parodies of human conflicts, like those of Jonathan Swift or Monty Python.

The baboon males of the group are the author's people. Sapolsky identifies with them in many ways, especially when females are concerned: he claims to have crushes on various study animals, including Bathsheba. The zoomorphism sometimes extends beyond the primate order: Sapolsky also claims that, after he has been in the bush for a long time, he gets turned on by the breasts of female elephants. Humans are animals, and animals seem human. If he walks the line in his scholarly writing, Sapolsky systematically blurs the line between species throughout this informal account, in which the ordinary rules do not apply and the heroic experiences of other primatologists are replicated in a carnivalesque mode.

In many ways, Sapolsky's hardships were predicted by Goodall's life at Gombe two decades earlier, and Fossey's at Karisoke, but his privations are played for laughs rather than heroics. The limited diets of Fossey and Goodall were forced upon them by circumstances; Sapolsky chooses, year after year, to cart out to the bush crates of execrable bony Taiwanese mackerel in tomato sauce because it is cheap and he doesn't want to take time to find something better. Once, he inadvertently colludes with poachers and joins them in their meal of barbecued zebra. His worst food mistake is buying tamarind, without tasting it first, as his only staple for a trip through the Turkana Desert, which straddles the border between Ethiopia and Kenya. He discovers that this legume is a strong, almost inedible blend of salty, sweet, bitter, and sour tastes; his supply is dried and therefore more intense. Sapolsky steals food when the money runs out, sleeps on the ground in a poor excuse for a tent, and is rarely clean. He suffers the isolation and loss of bodily control mentioned in almost every narrative written by field primatologists, but instead of the weight loss, fever, and dangerous diseases suffered by Goodall and Fossey, Sapolsky regales the reader with tales of foot fungus and diarrhea—once under the watchful, concerned eyes of female elephants, who gently wave their trunks as they consider his agonies. The carnivalesque element of the fragile, permeable, out-of-control body is everywhere in this book—screamingly funny at first, and terribly sad before it is all over.

Carnival is also the mode of pratfalls. From his first arrival in Nairobi, where he immediately falls victim to a con artist, to months or years later, when he finds himself at the wrong end of a Masai spear after insisting that the baboons are his cousins, Sapolsky seems always to be the butt of a practical joke. During a trip through the Sahel, he is forced to relieve himself under the beam of a flashlight and the eyes of an entire giggling village. On the same trip, narrated at the end of part 2, "The Subadult Years," he falls in with six Somalis, who, under the guise of friendship, force him into several sleepless, terrifying days and nights of driving through Sudan, playing cards, and consuming numberless Coca-Colas until, finally, in an altered state induced by sugar and caffeine, he escapes when their vehicle breaks down. Some of the dangers he encounters on this journey are natural: for example, voracious army ants, worthy of an appearance in a 1950s B movie, attack him in the middle of the night, after he has blown out the fire in his hut. Naturally, his screams are a source of fun for the villagers, who have thoughtfully lit the fire to repel the ants.

Not everyone Sapolsky meets presents an obstacle. Cassiano, a reverent animist who guides him through the forested mountains of Sudan, is nothing if not helpful. But the good guide is balanced by Sapolsky's guide through the Virungas—resentful, competitive, bullying, and, the author is convinced, homicidal. Even many of the genuine helpers on his quest are presented as comic figures. "Laurence of the Hyenas," whose research is conducted "on the next mountain," serves as a mentor and sounding board.[11] Laurence Frank—his real name, given only in the acknowledgments—has in fact made groundbreaking contributions in Sapolsky's own field, as well as new and startling discoveries about hyenas.[12] Still, the moniker, a comic echo of "Lawrence of Arabia," and the image of the scientist wearing cast-off army night-viewing goggles to sneak around a pack of hyenas, make him a comic figure. Just as one good guide is balanced by a very bad one, the kind mentor is balanced by an unkind one—Sapolsky's (unnamed) professor, who suggested and arranged for his field experience in Kenya, provided the initial funding, and then forgot to supply further funding, so that at one point Sapolsky is stranded penniless in Nairobi to live off his wits.

Sapolsky's Kenyan helpers, mostly dedicated and ambitious (in particular Hudson Oyaro of the Kisii tribe and Richard Kones of the Kipsigi tribe), are also a blend of the serious and comic. One, unfamiliar with the ways of the city, rides a hotel elevator in Nairobi all day; another, completely without idiomatic grace, cheerfully curses in English. One of Sapolsky's Masai helpers makes an "ice" with the dry ice Sapolsky keeps for his baboon blood samples—using cow's blood rather than the orange drink powder favored by Sapolsky himself.

IV

Not everything in the memoir is funny, however, and an episode exactly in the middle of it foreshadows the serious ending. As if to remind himself of the contrast between this real world and the green world he imagined during childhood trips to the American Museum of Natural History, Sapolsky relates a dream-like interlude set in an idealized Arcadia in the middle of violence and chaos—a last look, perhaps, at the optimism and innocence of youth. He passes through a remote golden-age mountain village, in which everyone is hospitable and peace loving, and fun is dressing up for a Saturday night dance. Birds, bushbuck, and monkeys share the forest. A few days later, when Sapolsky and his guide Cassiano finally reach the highest point in Sudan after a dangerous and grueling hike, Cassiano kneels "quietly, reverently," and places a small bird feather in the cairn at the summit. "At that moment," Sapolsky comments, "I deeply envied every animist his religion."[13]

A more marked shift in tone and story content takes place in chapter 21, at the end of part 3, "Tenuous Adulthood," and it signals that no amount of foolery can change this story into a happy one. This chapter comprises two parts, separated by white space. There is no logical or clear narrative link between the two parts. But in both Sapolsky shows himself beset by loss, grief, and rage.

The first part of the chapter is about the tensions between Sapolsky's fieldwork and his laboratory research, which come into sharp and painful focus at the discovery of an inexplicable baboon death. His grief is so extravagant that it startles several Masai women walking out on an errand; they are even more concerned when, on the way home, they find him carrying out a ritual burial of the corpse, which he has encircled with olives and figs, this particular (as yet unnamed) baboon's favorite foods while he was alive. Sapolsky seems to be as startled by his own reaction to the death as the Masai women. His grief over this one individual clearly violates the scientific objectivity he has always claimed for himself. The protocols of science in which he has been trained require valuing all the animals the same, rather than as individuals, and valuing human life so far above nonhuman primate life that an unnecessary animal death might be an unfortunate waste of a resource, but nothing more. Sapolsky is caught unaware by his grief over a favorite.

Then, when more animals die, the male dominance hierarchy fluctuates wildly, and the life of the entire study troop is disrupted. Sapolsky gradually realizes that his research, which consists of both fieldwork and laboratory work, involves ethical dilemmas not easily resolved. Back home, in his experimental research, things occasionally go wrong and animals are killed needlessly. At these times, Sapolsky writes, "I'd have dreams where I was Dr. Mengele—I'd wear a fresh new lab coat, and welcome the animals to their 'hotel,' the euphemistic nature of the word being discernible by them despite my Germanic accent."[14]

The conflict between the good for humans that comes from animal research and the evils involved in the process is not the only dilemma: in fact, as Goodall discovered, energy spent doing research must be subtracted from the energy available to protect free-living animals in their natural habitat. "Every primatologist I know is losing that battle," Sapolsky writes.[15] This first section of the chapter thus prefigures a discussion of layered moral failures in the book's last chapter, and with the ludicrous description of the author's grief, the burlesque elements of the memoir meet and overlap with the serious.

The second part of chapter 21 recounts Sapolsky's 1986 visit to the Virungas, only six months after Fossey's murder— "the stuff of legend," he calls her.[16] This section is another ragged edge in Sapolsky's memoir, where the burlesque and the serious mingle. Although the chronology is vague, he apparently undertakes the journey a few months after his discovery of the first baboon death. This whole chapter is about emotion, and Sapolsky describes his close-up encounter with the mountain gorillas in emotional terms, again abandoning scientific objectivity while being true to the impulse that propelled him into primatology in the first place:

I had a dream that summarized my feelings far better than I could when awake. It was a dream so tender, so ludicrously senti-

mental, so full of beliefs that I do not have when awake, that I still marvel at it. I dreamt that a certain brand of theology turned out to be true. I dreamt that God and angels and seraphs and devils all existed, in a very literal way, each with potential strengths and frailties much like our own. And I dreamt that the rain forests of the Mountains of the Moon were where god placed the occasional angel born with Down's syndrome.[17]

It would have been impossible for Sapolsky to describe an encounter with the mountain gorillas or a trek through their mountains without connecting virtually every moment of the experience with the memory of the woman who made the gorillas well known to the rest of the world. Sapolsky's impressions of Fossey are ambivalent: hero worship as an undergraduate (when he taped Adrienne Rich's poem "The Observer" above his desk), disappointment when Fossey brushed him off after a lecture, skepticism of her science as he became a scientist himself, continuous admiration of her fortitude and commitment, wonder at her discoveries, and, at last, grief. Another unwritten emotion, which seems evident in Sapolsky's juxtaposition of his own internal conflicts with Fossey's single-minded commitment, is a sense of inadequacy, and so his account of the visit to Karisoke is filled with tension. Remembering his feelings as he stood at the graveyard in Karisoke, he writes,

Fossey, Fossey, you cranky difficult, strong-arming self-destructive misanthrope, mediocre scientist, deceiver of earnest college students, probable cause of more deaths of the gorillas than if you had never set foot in Rwanda, Fossey, you pain-in-the-ass saint, I do not believe in prayers or souls, but I will pray for your soul, I will remember you for all of my days, in gratitude for that moment by the graves, when all I felt was the pure, cleansing sadness of returning home and finding nothing but ghosts.[18]

In this section, Sapolsky's anxiety of influence erupts overtly and fiercely—the only time such conflicted feelings about a woman role model are expressed in his work. He understands the allure of magical thinking and the difficult emotional work of overcoming it. Fossey did not take Sapolsky's youthful passion for the gorillas seriously, but that is perhaps not the real reason for his mixed feelings. In her logic-defying life and horrible death, she avoided the work of balancing the requirements of science with her love for the animals. Always involved in controversy and conflict, Fossey nevertheless was able to evade moral ambiguities.

Amid the myriad attacks by Fossey's critics—against her science, her approach to conservation, and her treatment of colleagues and the Rwandan people—perhaps Sapolsky has come the closest of all to identifying, albeit indirectly, the sources of animosity toward her within the primatology community. And in the end, his gratitude for her sacrifice wins out over other emotions.

V

Sapolsky is a self-conscious writer, not entirely at ease with his medium in *A Primate's Memoir*, as he is not at ease with the tensions inherent in primatology. In fact, as his comments about the lexicon of science suggest, his conflicts about writing and the tensions within his scientific practice overlap to a significant degree. Unless one adheres to a preestablished formula, writing things down in an authentic way is always a process of discovery—about one's subject and oneself. Such writing cannot be easy or comfortable, and Sapolsky's vexed position in relation to his material is a distinctive feature of primatology field narratives in general.

Unease characterizes Sapolsky's narration of disease in the baboon group, from the first shocking death and pathetic burial to his full understanding of the plague and the chaos that eventually decimates the group: "I have decided that it is time to tell how my baboons ended. I have tried throughout this book to give some attention to the style of writing, to try to shape some of these stories. Here I will not try. Things unfolded in an odd, unshaped way. There were villains, but they were not quite vile enough to satisfy. There was no show-down. These are not a crafted, balanced set of events, and the telling of them will not be particularly crafted either."[19]

The plague of bovine tuberculosis that infects the troop originates in a garbage dump behind a tourist lodge, into which the offal from infected meat is thrown, to be fought over and consumed by many baboons, including full-time residents. Sapolsky and several colleagues in Kenya quickly turn their attention to the diagnosis, etiology, and possible treatment of the outbreak. The necropsies and other tests carried out by these scientists reverse the carnivalesque rever-sals. The ravaged bodies of the baboons—laid open under the scientist's knife and to the view of the reader as well—are no longer comic but horrifying. The plague takes a dozen or more of Sapolsky's study animals, including his favorite, Benjamin, and "Manasseh, who died writhing in front of a laughing crowd of staffers at the lodge."[20]

More horrifying still are the results of the scientists' detec-tive work, which uncovers not only an infected population of local baboons but an infected human world. In a scam that is typical but more deadly than most of the scams Sapolsky has stumbled across during his years in Kenya, sick cows belonging to the Masai are being briskly sold to the local meat inspector, who in turn sells the beef to the tourist lodge. Sapolsky tries to involve government higher-ups in prosecut-ing this activity, which is both illegal and dangerous, but since bovine tuberculosis is more dangerous to baboons than to humans, and since it is not transmitted through consumption of infected meat, no one with the power to stop the abuses is interested in doing so. As the responsible scientist, Sapolsky publishes his findings about the baboons' susceptibility to this particular strain of tuberculosis in professional journals, knowing that no one who can intervene in the situation actually reads scientific articles. Finally, he realizes that his fantasies about asking the British royal family to intervene will come to naught, even though some of them frequent the lodge. Down the causal chain, here are the rich tourists, too, implicated in Kenya's social and economic ills.

The plague runs its course, and the baboons are not the only ones affected by it. "I am a different person now and at a different point in life than when I started here," writes Sapolsky in the last chapter. "Once I was twenty and I feared nothing but buffalo and I came here to adventure and to exult and to defeat my depressions, and I had an infinity of love to expend on a troop of baboons."[21] The author's internal conflicts are not resolved, but they have faded with time. At last, years after these events, Sapolsky returns with his wife, Lisa, to visit the troop. He catches sight of Joshua, who resisted the lure of the tainted meat. Joshua has grown feeble and a little senile but remembers his human friend. And then "Lisa and I did something rather unprofessional, but we didn't care. We sat down next to Joshua and fed him some cookies. English digestive biscuits. We ate some too. He went about it slowly, grasping the end of each delicately with his broken old fingers, chewing with small, fussy toothless bites, continuing to fart occasionally. We all sat in the sun, warming ourselves, eating cookies, watching the giraffes and the clouds."[22] With uninhibited kindness and fellow feeling, carnival is no longer necessary, no longer violent, but traces of carnival remain in the chewing, farting, and shared openness to clouds and sunshine.

The baboon troop, too, has changed.[23] It is more tightly knit, less aggressive, more cooperative. Other field researchers have shown that behavior in different baboon populations can be remarkably varied and flexible, so this change may be permanent, and it may reflect both a change of "rules" and a genetic change due to the recent reproductive success of the "nice guys." As in Sapolsky's representation of his human community, the females are steady and self-contained, and they mostly read the world aright.

As primatologists continually discover more evidence for primate individuality and more behavioral variations among and within species, the search for a "primate pattern" no longer animates primatology research. But it is still valid to contemplate a baboon troop as a comic microcosm. It is still helpful to consider the experiences of one man, living in the wilderness (a compromised wilderness, to be sure) with his study animals, as a broad outline for the lives of other field primatologists. Laughter is part of that broad outline, even if it cannot be sustained.

And for some quieter laughter, we may turn, at last and in conclusion, to some stories about primates far less known and understood—and far more at risk in this perilous world—along with the scientists who observe them in likely and unlikely places.

Conclusion

We will more and more be studying nonhuman primates as affected by human primates. Willy-nilly, we will make a virtue of necessity. We will (and have) linked conservation biology with small group genetics and viability. But also on the behavioral side we will study migration between isolated forests, the effects of sparse or highly crowded populations, edge effects. And then claim that probably a lot of natural selection has always taken place in extreme circumstances. . . . This is all part of the general ecological shift from believing in stability to being impressed by patchiness, change, catastrophe, at different scales. Looking at primate social behavior in the face of human-induced change will soon be respectable. . . . I have hardly ever seen a primate field site that wasn't massively impacted by humans . . .

—ALISON JOLLY, quoted in
Strum and Fedigan, *Primate Encounters*

I

Remembering his two most influential teachers, Niko Tinbergen and Konrad Lorenz, the baboon researcher Hans Kummer proposes that "the two great ethologists represent the two ancient forms of the animal-human relationship. One tamed the animals like a herdsman, the other spied on them like a hunter."[1] In this book, I have thus far concentrated on the narrative strategies of Schaller, Goodall, Fossey, Strum, Hrdy, Smuts, Strier, and Galdikas—hunters all, like Tinbergen, who spied on apes and monkeys in their natural environments and clearly distinguished their own work from that of laboratory scientists. I have also offered a reading of *A Primate's Memoir* by Sapolsky, who studied baboons in both the field and the laboratory and experienced a painful tension between the two forces at work in his scientific practice.

To characterize Sapolsky's research as a combination of herding and hunting does not begin to reflect these tensions. The stresses in his work were exacerbated by field conditions that would have been almost unrecognizable even a generation earlier. Sooner or later, the other primatologists whose stories I examined earlier in this book also discovered that the hunter model, however emotionally and scientifically satisfying, was not, by itself, sustainable in the late twentieth century. It is becoming less so in the twenty-first.

Now, as primate habitat shrinks to a damaged sliver of what it was merely fifty years ago, the hunter's role merges with the herdsman's (or the shepherd's, in the case of scientists who work with rescued animals). Since the turn of the century, even the stories have changed along with the practice of primatology. Romance, tragedy, comedy, and the novel are giving way to other narrative forms. The distinction between forest, sanctuary, reserve, and laboratory blurs as human encroachment into woodland and savannah has put at risk almost every primate on the planet, including ourselves. Field primatologists are, in a sense, starting over. Even though the shapes of the stories are still fluid, new themes are emerging. Their power remains to be seen.

II

"He's asking permission to climb onto that branch with her," Monica Mogilewsky says with a chuckle. "I like it that they're female dominant." We are watching a pair of red-ruffed lemurs, housed together in a roomy inside-outside cage at the edge of a thirteen-acre forest in southwest Florida.[2] The pavilions at the Myakka City Lemur Reserve are devoted to "special" lemurs—breeding pairs, individuals who don't work and play well with others, and everybody else during times of environmental stress. Wildfires and hurricanes are routine in this part of the country.

Most of the time, though, the majority of the forty lemurs at this breeding and research facility wander freely through two patches of woodland—one an oak and pine hammock, where we're standing, the other mostly slash pine. The lemurs who hang out in these woods are provisioned, but they are able to find about half of their food on their own. Of course, it's impossible to reproduce the Madagascar habitat in which lemurs evolved, but the Lemur Conservation Foundation tries to make the animals' new space as naturalistic as possible. "There's even predation!" Monica tells me. "One of them went missing, and we think a hawk got him." The forest and the laboratory here are one and the same.

Most of the prosimian residents cared for by the Lemur Conservation Foundation came here more than a decade ago from an overflow at Duke University's National Primate Center, and some have been born at the reserve, the successful outcome of a captive breeding program intended eventually to bolster the numbers and genetic diversity of free lemur populations in Madagascar. Since brown lemurs offer the best long-term prospects, they are a special focus of the breeding program. A few residents of the reserve are refugees from the pet trade—a topic likely to unite the most aggressively self-protective experimental primatologist with the most ardent PETA activist, because primate pets are usually dangerous to their human companions and often desperately unhappy. The lemurs at the Myakka City reserve are tame, but tame is not the same as domesticated. Monica is eloquent on this topic because one of the female brown lemurs at the center, a breeder from the pet trade, has had infants stolen from her so often that she now carries her young until they are as big as she is, and she can be vicious when she feels threatened. Most of the lemurs are friendly, though.

Monica belongs to a new breed of primatologist, apparently untroubled by accusations of anthropomorphism, able to navigate outside the language of behaviorism, and willing to extrapolate (in conversation, at least) lemur emotions and perceptions from observing them and triangulating with human feelings. She has also carried out her own experiments and discovered higher cognitive levels than most people ascribe to lemurs. At another cage, Monica suggests that I lay my palm on the mesh so a friendly male can investigate. He licks my hand with gentle, exploratory strokes. "I think they like us," she says, "but they don't think we're very smart because we can't communicate with them as well as they communicate with each other." In any case, the lemurs find humans useful. Outside a mesh cage in the heart of the oak and pine forest—the new home of a rare eastern lesser bamboo lemur—a ringtail takes a running leap onto Monica's shoulder before landing on top of the bamboo eater's cage (see fig. 4). This might be one of the males I saw a little earlier, lined up on a live oak limb to observe the humans as we tracked them down.

In the smaller slash pine and wetland enclosure, we find another lemur group, red-ruffs, assembling on a log to watch us watching them. They are bright eyed, calm, and curious. Who is this unknown human? Anyway, two can stare as well as one. Now I always think of lemurs in a line, watching me with calm curiosity. Even though they are not congenial as pets, it is no wonder that people want to own them.

Literally, the word *lemur* is Latin for ghost. Gonzo journalist and novelist William S. Burroughs, who became fascinated with lemurs in his later years and visited the Duke research center for a face-to-face encounter, wrote in his short novel

Fig. 4 Monica Mogilewsky with a ringtail lemur companion,
at the Lemur Conservation Foundation in Myakka City, Florida.
Photo: Katie Chapman.

Ghost of Chance that "they gambol, leap, and whisk away into the remote past before the arrival of man on this island, before the appearance of man on earth, before the beginning of time."[3] Lemurs inspire nostalgia in science writers as well. In *The Song of the Dodo*, a searching examination of trends in biogeography and extinction, David Quammen describes lemurs, especially the rare and charismatic singing indri, as if they were already lost. Indri have large throat pouches, he writes, perhaps an adaptation for making their remarkable music.

Otherwise, the indri's anatomy and physiology are poorly known; few scientists have laid hands on a dead specimen, let alone a living one. It is clear, though, that the indri's slow rate of repro-duction and its low population density make the species especially vulnerable to extinction. Probably its body size does too. . . . The dozen or so species of lemur that have gone extinct since humans colonized Madagascar were all large-bodied animals, generally much bigger than the lemur species that have survived. If that trend holds, the indri will be next.[4]

The tone of Quammen's paean to the lemurs of Madagas-car may have as much to do with the far distant past as the future. All primates share anatomical features—a domed cranium, relatively large brains and generous cerebral capabil-ity, forward-facing eyes and ears, hands and fingernails, thumbs, and free limbs—that is, arms and legs that can move in multiple directions because they are positioned completely outside the torso. The majority of primates, including humans and most lemurs, also have complex and long-lasting social relationships. In addition, all primates have a common his-tory—up to a point. Ancestral lemurs were separated from other primates several millennia after Madagascar broke off from eastern Africa about 120 million years ago and slowly drifted away. Perhaps the ancestors of modern lemurs migrated from Africa by means of shallow water or flotsam, or took a more circuitous route through Europe and the Indian subcontinent, to which Madagascar remained attached for many more millennia. No one really knows. In any case, lemurs evolved nowhere else.[5]

Most scientists estimate that lemurs have existed for forty to fifty million years. A recent fossil find in Germany, estimated to be about forty-seven million years old and nicknamed "Ida," may prove to be a missing link between the simian and prosimian branches of the primate order. Ida is more similar to modern lemurs than to any other living species, but her ankles and teeth, in particular, reveal a kinship with ancestral apes and monkeys, too. Like finding a distant cousin in a foreign land or discovering a cache of letters from a great-great-grandmother, claiming kin with lemurs through Ida involves an imaginative leap into a dream world of long ago and far away.[6]

But the nostalgia also bleeds out of real concerns for the lemurs' future. Instead of adapting to threats from large predators and resource competition from many other kinds of herbivores, these animals competed among themselves, evolving sympatrically (in the same place, but with different ways to make a living) to fill the available niches on Madagascar. A good illustration of sympatric lemur evolution is the existence of three distinct species of bamboo lemur in the same very small area of the island: the golden bamboo lemur, who eats the cyanide-laced roots and shoots of the giant bamboo; the broad-nosed gentle lemur (or greater bamboo lemur), whose mouth is adapted to break open giant bamboo stalks for the pith; and the gentle bamboo lemur, who eats mature bamboo leaves, but usually not leaves of the giant bamboo. Under natural environmental pressures such as storms and fires, generalists such as humans and savannah baboons fare better than specialists such as the bamboo-eating lemurs. Under abnormal pressures such as habitat destruction and fragmentation, hunting, and global warming, tiny populations of specialists may be doomed to extinction if extraordinary measures are not taken to preserve them. That is why the lemur conservation program in Myakka City, Florida, has begun to focus on the breeding of common brown lemurs, the least specialized of the suborder *Lemuroidea*. This species could last for a very long time if it had to be reintroduced to Madagascar.

Great environmental pressures on lemuriform species in Madagascar began within historical time: humans migrated onto the island only about two thousand years ago. Within a few hundred years, elephant birds (the roc of fairy-tale fame), two giant tortoise species, and fourteen species of lemurs (including a human-sized species) were driven to extinction, probably by hunting. Unlike almost all other primate species, Malagasy lemurs have never been systematically studied as members of undisturbed ecological communities. So, for the scientists who love them, lemurs have always already been a dying branch of the great primate order to which we humans also belong.

III

The lemur reserve in Myakka City is an ark. Time is short, and arks represent extreme, last-ditch efforts to save imperiled animals in case all else fails, or is liable to fail. Any legitimate, regulated captive breeding program for rare or endangered animals can be considered an ark, not only in reserves such as the Florida facility but also at research centers such as Duke's National Primate Center, although breeding is not the primary mission of most research centers.

Reserves are also islands. As Jane Goodall noted, Gombe has become an island, an isolated patch of ground out of which the chimpanzees cannot travel, and into which chimpanzee visitors cannot enter. (Fortunately, with the support of the local community in Tanzania, the island is at this moment a little larger than it was, since more trees have been allowed to flourish around and outside the park perimeter in recent years.) The home of the mountain gorillas in the Virungas of central Africa is also an island, isolated by cultivation and, according to a 1995 *National Geographic* article by George Schaller, soil hardened to rock-like density by the feet of war refugees.[7] In 2008, Mark Jenkins reported in *National Geographic* the outright murders of seven mountain gorillas, probably by illegal charcoal traders. Brent Stirton's sensitive photography of the corpses and the grief-stricken park guards shows the extent to which the forest has shrunk in relation to the matrix—the land around it.[8] All too often, the local people are overlooked in primatology narratives, but Stirton's photography shows their investment better than words could do. Unfortunately, they are no more able to stop the violence against the forests and their inhabitants than anyone else.

Likewise, though without the violence, South American muriqui populations (as well as other endangered primate species such as the golden lion tamarin) are now confined by logging and agriculture to several isolated areas within what was formerly a vast coastal forest.[9] Orangutans have evolved on the large islands of Borneo and Sumatra, and the areas they inhabit have been steadily cut away, leaving much smaller islands of habitat. In fact, almost all free-living nonhuman primates now inhabit islands of various kinds. An obvious problem arising from isolation is genetic viability: too small a population results in inbreeding and sometimes an unbalanced male-female ratio, both of which can hasten a population or even a species toward the vanishing point.

Yet genetic viability is only part of the problem. No species at risk can be saved where it evolved unless it is considered as an integral part of a biological community. Thus, primate conservation cannot succeed without ecological data and analysis.

If primatologists who began their work during the era of Leakey's "trimates" have understood the challenges in political terms, they have not been incorrect. However, by the end of the twentieth century, ecology, a field that developed along a track parallel to primatology, had supplied an additional perspective in environmental discussions, and primatologists began to carry out their conservation efforts in more convincing and effective ways as a result of emerging ecological theories.

IV

Human language is linear. Human beings experience their lives in linear fashion. In the stories that inspire human feelings and actions, narrative—that is, linear form—is strongly marked. At the core of human consciousness, "narrative curiosity," so called by English novelist A. S. Byatt, may rival the emotions of love and fear.[10] Appealing to narrative curiosity or greed is, as I have tried to show in this book, one of the best ways for a scientist to educate the public about the kinship between human beings and other living things, and about our responsibilities for those other lives.

Fragmentation science addresses the dynamic interaction among three zones or landscape types: the core protected fragment (with due attention to ecological relationships within it); the edge or perimeter of this area; and the matrix (the larger area surrounding the protected fragment).[11] Knowing the ecological history of a fragment can also be crucial in managing it for the future, and if that small piece of history is woven into the larger tapestry depicting a species, a people, and a nation, it becomes a compelling story. In looking to the future, then, narrative practices can be doubly important for primatology.

A double narrative interest—landscape history and human history—runs throughout the lifelong project of Alison Jolly, a lemur specialist and author of *Lords and Lemurs: Mad Scien-*

tists, Kings with Spears, and the Survival of Diversity in Madagascar (2004). Jolly was one of the first primatologists to devote her career to the study of prosimians in Madagascar and one of the first generation of women primatologists to make her mark in the discipline. Here is one example of her findings: Jolly's observations of the relative lack of sexual dimorphism in lemurs, along with female dominance in most lemuroid species, suggested to her that male dominance across the primate order might be a coincidental effect of sexual selection. In other words, if large size is adaptive for male apes and monkeys fighting over females, or trying to impress them, then the winning males can also parlay this advantage into bullying their hard-won females. Jolly was pretty sure larger body size didn't evolve so that males could control females. Any feminist alive would say that this insight was common sense, but Jolly was the first primatologist to theorize the subject in this way.

In addition to a full slate of scientific texts, Jolly has also written natural histories of Madagascar (complete with the destructive human context) for the general public, children's books about prosimians (in cooperation with the Durrell Wildlife Conservation Trust), articles for *National Geographic,* and *Lucy's Legacy* (1999), a meditation on evolution, sexuality, intelligence, and the human impact on the biosphere. As the title suggests, *Lords and Lemurs* is about how political power, traditional culture, and science converge in the issue of protecting animals and biological diversity.

There are two plots in this book, one of which takes place in evolutionary time, the other in historical time. The lemurs' plot, of course, takes place within evolutionary time. The historical plot begins with the settlement of Madagascar, probably less than two thousand years ago, by African and southeast Asian people. Highlanders from Asia exploited African settlers in the south and west of the island, and this pattern has continued to the present day. This rift within the politics of the island grew deeper after the arrival of Europeans in the seventeenth century and has become still more complicated with postcolonial pressures in the wake of the French divestment following World War II.

Another pattern has operated in Madagascar throughout human history on the island. Environmental pressures, which long ago resulted from widespread deforestation and all its attendant ills, inspired an early monarch to protect the forests, and in 1927 Madagascar became the first African country to reserve large areas of its forest. But now the landscape is so degraded that scientists and conservationists who have worked there (such as Penelope Bodry-Sanders, founder of the Lemur Conservation Foundation) report their sadness and shock at flying over the island for the first time. As deforestation proceeds at lightning pace, the land visibly bleeds into the rivers. But wildlife and conservation organizations have treated the island with the utmost concern, and the Malagasy people themselves have been forward-thinking on environmental issues. Ironically, both overuse of the land and deep concern for the island's ecosystems are part of their history.

Jolly's story focuses on Berenty, a tiny reserve on the Mandare River, at the southern tip of Madagascar. This forest has been a fragment since it was set aside for protection in 1936 by a French colonial landlord, Henry de Heaulme. At same time, the de Heaulme family developed a sisal plantation in the dry spiny forest on the other side of the river, preserving ten-meter strips of the original vegetation throughout to prevent wind erosion. The numerous Tandroy people who live in the region work the sisal plantation and factory, while preserving many features of their traditional village life. During tough economic times in the 1990s, the de Heaulme family established an ecotourism concession that now generates enough income to keep the sisal operation going even in times of low demand for its products, thereby protecting the reserve and supporting the human population that might otherwise put more pressure on the forest. Most of the tourist staff are Tandroy, and many Malagasy scientists conduct their research at Berenty along with Westerners.

Berenty is a model for many more recent *in situ* conservation projects worldwide. Even with the support of local people, official Madagascar, the scientific community, and ecotourism, however, the forest shrinks bit by bit, as rainfall declines, forest edges slowly recede, and the river dries up. There are no more crocodiles and eels, which historically made the river a dangerous source of good eating. Climate and weather patterns are changing. Under current political and economic policy, there are simply too many people for the island to support. And as in most places where the population exceeds carrying capacity, Madagascar has been beset by political and even military turmoil.

Jolly begins and ends her story with the lemurs, whose individual lives are timeless and cyclical, but whose species life is now subject to human history and the vagaries of human nature. Even the lemur demographics of Berenty are not typical of undisturbed habitat. The original population of the reserve included ringtails, lepilemurs (small nocturnal leaf eaters), tiny mouse lemurs, and the singing, leaping sifakas. The de Heaulme family accidentally released brown lemurs from the west coast and later added browns from the eastern rainforest who had been rescued from a pet market. These two genetically distinct subspecies have somehow, to everyone's surprise, cross-bred and continued to reproduce. Their progeny would be unwelcome in zoos but are interesting to scientists who work in fragments and investigate free-living hybrid populations.

Two lemur characters in Jolly's story are Fan, a ringtail, and Cream Puff, a brown, the alphas in their respective families. When threatened, the lemurs can be nasty fighters in both internecine battles and interspecies wars. Jolly clearly prefers the striking and orderly ringtails, but she admits grudging admiration for feisty, chaotic, and more generalist brown lemurs: Cream Puff may be a serial killer of ringtail infants, but she's a good mother to her own! The lemur tribes are introduced in the first chapter as denizens of the de Heaulme domain—lounging in backyards, strolling through village centers, and leaping through forest trees. Their lives and needs punctuate Jolly's layered narratives of the de Heaulme family, the neighboring Tandroy, and the whole nation. Like the human beings in this story, the lemurs survive against the odds. Fan and Cream Puff and their families manage for decades to survive their own fights, droughts, overcrowding, and inappropriate banana gifts from ignorant visitors.

Integrated with the human story, the lemur chronicle differs from it fundamentally, and Jolly emphasizes this difference. We modern humans live separated from the cyclical time of our ancient primate ancestors (human and otherwise); we are conscious of living in a line, not a spiral. But watching the lemurs, even secondhand through the eyes of a scientist, encourages a twenty-first-century reader to retrace those first steps our ancestors made, half a million years ago, into the present. At the end of the book, Fan, Cream Puff, and their descendants continue to delight tourists, serve the desires of scientists, live their own lives, and propagate their kind:

Each day is a day. In the bitter season when their babies are born, they claim space as their own, each in her own way. . . . They nurse their babies from their own energy and from the season-end fruit of the tamarind trees.

In mid-September the tamarinds turn yellow-gold. . . . Tiny yellow tamarind leaves fill the air, swirling as they fall. Now the branches are bare for two or three weeks.

Tree by tree, and soon all over the reserve, new leaves peek through. . . . At last the lemurs eat their fill of protein-rich rose-pink leaves. . . . Troop warfare slows. There is enough for all.

Month-old babies totter off their mothers. They begin to play at hop-and-pop all through siesta time, while their parents doze and groom. They do not look to past or future, only to the warm afternoon, their playmates, and their mothers' milk, in the enchanted forest.[12]

V

In contrast to the elegiac note that ends *Lords and Lemurs*, *Bonobo Handshake* by Vanessa Woods chronicles the ways in which bonobos have been thrust from their own cyclical, evolutionary time into historical time. Woods, who began her professional life as a journalist specializing in wildlife, has since become a research scientist in environmental anthropology at Duke University. With her husband, Brian Hare, she has begun research on dogs as well as primates—not an unusual track for primatologists (Donna Haraway and Barbara Smuts have followed the same route).

Woods's first primatology adventure, narrated in *It's Every Monkey for Themselves: A True Story of Sex, Love, and Lies in the Jungle* (2007), resembles in some ways the stories considered earlier in this book. Here, Woods chronicles her adventures with a group of field scientists as they follow capuchin monkeys in Costa Rica—a grueling affair of brutally early risings, surly companions, stinging vines, falls, near-lethal bee stings, and frustration. Like Sapolsky before her, Woods characterizes herself as a joker—or, more often, the butt of jokes. She continually makes unfortunate decisions (too much luggage) and bad judgments (about her housemates) similar to Sapolsky's, and, in addition, she allows herself the folly of risky sexual encounters. Her greatest consolation is playing the guitar and singing to the street dogs who have been adopted by residents of the "monkey house." The dogs seem to appreciate the effort.

Woods's companions are not loners, either. They are packed into a house in a small village, living in each other's pockets and replicating the competitive behaviors they observe in their study animals to a disconcerting degree. Although the monkeys are still wild and sometimes difficult to find in the Costa Rican jungle, the "field" is crowded with the humans who have come to study them.

Woods maintains an eccentric persona in her next book, too, written about her experiences with chimpanzees in Uganda and bonobos in the Democratic Republic of Congo. Again, she presents herself as childish, petulant, distractible, and oversexed. One weave in *Bonobo Handshake* (2010) is her relationship with Hare, whose background as a technician at Yerkes Institute outside Atlanta has given him both the insight and the skills to embark on innovative research methods and themes in a degree program supervised by Richard Wrangham. The lovers quarrel about everything, from returning for research to the DRC, one of the most dangerous places on earth, to Brian's insistence that she touch the bonobo Kikongo's penis in order to persuade him to finish an experiment, to Brian's comic infatuation with Malou, one of the female bonobo "toddlers," who has a crush on him.

Aside from the comic relief provided by Woods's jokes at her own expense, her adventures in *Bonobo Handshake* are more dangerous and potentially significant than her time in Costa Rica, and they lead, gradually, to a greater interest in understanding these animals in scientific terms. Woods literally transforms herself from a journalist into a scientist. Assisting Hare in his comparative chimpanzee and bonobo cognition experiments, she begins to design her own experiments, contributing to the scientific as well as popular literature about wild and captive primates. *Bonobo Handshake* signals a new mission for primatologists, a new figuration of the scientific field, and a profound examination of the relationships between humans and our primate kin, not just from the depths of our joined evolutionary past or psychosocial similarities, but in our political present and precarious future.

As Jolly predicted, the sanctuary, like the reserve, is becoming the field. In one of Woods's first conversations with him, Hare explains that funding for sanctuary work can enhance research facilities in sanctuaries, which offer better (more natural) conditions for research than many of the primate laboratories in the industrialized world. Hare observes that primate laboratories still sometimes rely on cruel, artificial, and outmoded treatment of and surroundings for the study animals, as well as torturous invasive procedures. Sanctuaries are, for the most part, real sanctuaries.

After meeting at the Ngamba Island Chimpanzee Sanctuary in Uganda in 1999, Woods and Hare move on to Lola ya Bonobo in the DRC to conduct the same experiments with bonobos that Hare has already tried with the chimpanzees under similar conditions. However, the bonobo sanctuary is not a sanctuary in the ordinary sense of the word. Despite its name, which means "Bonobo Paradise," Lola ya Bonobo is only a small island of relative peace in the midst of conflict, which intrudes into the sanctuary on a daily basis, along with the human caretakers who live just outside and the steady stream of infant and juvenile apes suffering from physical disease, grief, and post-traumatic stress syndrome.

The troubled DRC is located on the southern side of the Congo River; on the northern side is the similarly named Republic of Congo. (Sometimes they are referred to together as "the Congos.") An accident of evolutionary geography isolated one branch of the chimpanzee genus to the area south of the river, within an enormous oxbow stretching about four hundred miles from east to west. These apes developed into bonobos, previously known as pygmy chimpanzees because they are more gracile than their cousins the common chimpanzees. We humans are the third scion from the chimpanzee family tree. In genetic (and evidently behavioral) terms, humans, bonobos, and chimpanzees are equally related, sharing over 98 percent of one another's genes.

Numerically, the bonobo population is miniscule, and there is only one long-term field site, established in Wamba in 1973 by Takayoshi Kano. As Woods notes, "The Japanese have told us almost everything we know about bonobos: their diet, their habitat, their social structure."[13] Because there are so few of these apes, and because they live in areas so out of the way for outsiders, she adds, the bonobos have never had a Goodall or a Fossey to advocate for them in the West. Unfortunately, the bonobos' isolation does not protect them, as generations of civil wars and cross-border conflicts among Congo, the DCR, Uganda, and Rwanda have resulted in chaotic conditions that make enforcing laws against the bushmeat trade difficult or impossible. According to Woods, up to 80 percent of the meat consumed in central Africa comes from this trade, which relies principally on large animals because they are more convenient for butchering.

When bonobo parents are slaughtered, infants and toddlers too small to be sold as meat are offered in the marketplace as pets. In recent years, President Kabila's law enforcement officials have been willing to confiscate these tiny creatures, who are typically traumatized physically and psychologically. The little bonobos are then turned over to Lola ya Bonobo, located outside Kinshasa. The sanctuary was founded by Claudine André and has rehabilitated dozens of these lost souls. In 2009, it released several carefully chosen individuals into a remote forest reserve, or, in the phrasing of fragmentation scientist Laura Marsh, a "wild zoo."[14] There, they are protected by the local Pô people.

Fossey was murdered before Rwanda erupted in genocidal conflict, and although Goodall's graduate students were kidnapped in a cross-border raid from war-torn Uganda, she did not see firsthand the results of the war on the Ugandan people. In contrast, Woods has witnessed the consequences of war up close. Slaughter, privation, the destruction of habitat for humans and other animals, and rape seem ever present as results and weapons of war. Woods has known coworkers with limbs lost due to torture and others who have been psychologically devastated by less tangible losses, or even by their own participation in the violence. She writes of a group of women who, fleeing from their village, were overtaken on the road by a group of soldiers, then killed and mutilated for trophy body parts—and of a sole survivor who watched her daughters slaughtered and eaten before she, too, died, a short time after telling her story.

So the tiny bonobos who are brought to Lola ya Bonobo, sometimes too sad to live, are only a few among many members of our own order to suffer. As a result, Woods's story must be told in a new way. Biography, of course, is one thread of the story; by this time it is a familiar feature of the primatology field narrative. As Woods's individual war stories indicate, another thread is composed of the voices of war victims who tell her about their experiences. Still another thread is a thoroughly researched historical backstory, in which Woods traces the origins of these horrifying conflicts through the transatlantic slave trade, the subsequent rise of European colonialism, and the establishment of national boundaries in Africa on the basis of European political ambitions rather than preexisting ethnic boundaries. Ethnic conflicts within one country's borders—Rwanda, in this case—inevitably spill over to involve its neighbors, as refugees turn forests into deserts and displaced power brokers turn their energies abroad in order to recover privilege at home—in this case, toward the DRC. Extractive logging, mining, and forestry in the postcolonial period have exacerbated economic, national, and ethnic tensions still more. Wars in central Africa have been both funded and driven by industrialized nations' demand for the area's vast mineral wealth, including coltan, which is required in the manufacture of most digital devices.

If you are reading this book, you are probably implicated, as I am and as Woods observes that she herself is, in the civil wars of central Africa, the devastation of its wildlife, and the suffering of its people.

The happiest of the narrative threads in *Bonobo Handshake*, of course, is the story of what happens within the walls of the sanctuary itself, including anecdotes of happy and hilarious rehabilitated bonobos and the valuable research Hare and Woods conduct with them. This research reveals, like Frans de Waal's zoo studies, that bonobos and chimpanzees have equal levels of intelligence but vastly different social and emotional lives, which affect their ability to cooperate. If chimpanzees are driven by the food rewards and impatient with sharing, bonobos are scarcely tempted by the bit of fruit they are offered, and they are often distracted from the task at hand by the social interactions that surround it. The reward, such as the handjob Kikongo demands, has to be appealing. For bonobos, who have been dubbed "hippies of the forest" in the popular media, sex really is a big part of it, and Woods speculates that their cheerful pansexualism more often functions as a psychological release from tension than as play or a reproductive urge. In this respect, chimpanzees and bonobos could scarcely be more at variance!

As humans, we're a little bit bonobo and a little bit chimpanzee. As the world becomes more connected, the environment more degraded, and the need for resources more draining on the planet—in conjunction with the exponential and probably unsustainable growth of the human population—it behooves us to remember the doubleness of our ape inheritance and the wildly different behaviors of our closest kin.

VI

Lemurs and bonobos do not look to the past or future, but humans must, as part of our evolutionary investment in history, language, and stories, not to mention our survival on the planet. If Darwin set us on the course of telling stories about primates, Goodall and Schaller were able to tell stories about our primate kin where they evolved, in their own fascinating social arrangements. Fossey sounded the alarm when she saw these wild places and free animals under attack; thus, Goodall's romance gave way to tragedy. Strum and Hrdy wrote of monkeys and humans as neighbors—not always good neighbors, but at least neighbors who could survive in proximity to one another. But Strum's novel was succeeded by Sapolsky's parody, and Hrdy's postmodern search for literary form was followed by Galdikas's colossal confessional assurance. Neighbors are not always friends. Still, Jolly described one peaceful lemur-human neighborhood—in the midst of an ecological war zone.

For Woods, though, that war zone is both political and ecological. By her own admission, the sanctuary bonobos may not behave as they would in their natural habitat, and even though she ends with the hope that some of the bonobos can survive once more in the forest, on their own, that possibility remains to be seen. Meanwhile, the population of bonobos in their natural habitat, untouched by human conflict, thins out into a dangerous edge of near extinction.

In the accounts of Jolly and Woods, the continuous narrative gives way to complex weavings of history, political commentary, personal experience, and natural history. And the thick description of the early narratives—in fact, the stuff of scientific data—gives way to descriptions of degraded or lost habitat, human misery, and human attempts to improve the lives of all primates, ourselves and our ape, monkey, and lemuroid kin. Inevitably, the animals are not so much the gravitational center of these later books. If the narrative excitement remains vivid in such narratives, the thick description, which gives earlier field narratives the texture of romances, tragedies, confessions, and novels about nonhuman primates, is diminished in proportion to the intrusion of human interests and human violence.

Species are dying out before they have been scientifically described, and known fields are vanishing so quickly that new discoveries about primates must be made under new and still-evolving conditions, not through traditional fieldwork. Are the stories dying out as well? In *The Animal Connection*, Pat Shipman argues on the basis of the paleontological and cultural record that humans evolved within a context of close associations with other species. Without those other animals as companions, adversaries, and mirrors, and without the stories that can only be written if they remain themselves, who are we human beings, after all? If researchers can find ways to create new narratives about primates in sanctuaries and wild zoos, perhaps we humans can once again migrate to the periphery of the stories we tell about our primate kin. Only then will the mirrors offer true reflections. And maybe, just maybe, some of these fascinating animals can be left protected where they are in the forest, the jungle, the savannah, just as they have lived for millennia, not for our purposes, but for their own.

NOTES

Introduction

1. Goodall, *Hope for Animals and Their World*. Goodall mentions discoveries of the blond capuchin, the kipunji, six new marmoset species, and two new species of titi monkeys. All these species have been discovered partly because of primatologists' hard work, but also partly because deforestation has made monkey habitats more accessible. Goodall's information about new primate discoveries is echoed frequently in cover articles of the *American Journal of Primatology* and the *International Journal of Primatology*.

2. For a good summary of primatology history, see the first chapter of *Primate Encounters: Models of Science, Gender, and Society*, edited by Shirley Strum and Linda Fedigan. Strum and Fedigan are the authors of this chapter, which serves as an introduction to the anthology.

3. See *Primates in Fragments: Ecology in Conservation*, edited by Laura K. Marsh. A pioneering anthology in the field of fragmentation science from the perspective of primatology, this volume contains numerous references to the contributions that primates make toward the general ecological balance of their habitat.

4. Diamond, *Third Chimpanzee*.

5. McEwan, "Evolution and Literary Theory," 11.

6. Wilson, "Foreword from the Scientific Side," ix.

7. I have consulted Darwin's *The Origin of Species* (6th ed.), *The Descent of Man, and Selection in Relation to Sex*, and *The Expression of the Emotions in Man and Animals*.

8. For information about foundational research in primatology, see Yerkes and Yerkes, *Great Apes*. For an extended discussion of Yerkes's contributions to the field, see also Donna Haraway's account of the early years of the discipline in "Monkeys and Monopoly Capitalism: Primatology Before World War II," the first part of *Primate Visions*, 17–111.

9. Bakhtin's most influential narrative theory (outside Russia) appears in *The Dialogic Imagination: Four Essays*.

10. Kuhn, *Structure of Scientific Revolutions*.

11. Heise, *Sense of Place and Sense of Planet*, 138.

Chapter 1

1. Poe, "Murders in the Rue Morgue," 397. The implications of Poe's scientific source are discussed at length in *Romancing the Shadow: Poe and Race*, edited by J. Gerald Kennedy and Liliane Weissberg; especially helpful are the articles by John Carlos Rowe (75–105), Lindon Barrett (157–76), and Elise Lemire (177–204).

2. Kennedy and Weissberg, *Romancing the Shadow*, 416.

3. Ibid., 422.

4. Ibid., 430.

5. Ibid., 424.

6. Huxley, *Man's Place in Nature*, 61. I have consulted the American edition that Huxley supervised in 1896.

7. Ibid., 77–78.

8. Darwin, *Descent of Man*, 1:191. Unless otherwise noted, references are to the reprinted facsimile edition of the first edition; the two volumes are bound as one.

9. In *Darwin Loves You*, George Levine summarizes the arguments of sociobiologists and social constructivists. Barbara Hernnstein Smith goes into detail about the history of constructivist understandings of knowledge in *Scandalous Knowledge*; evolutionary psychology is the subject of the last two chapters of this book.

10. One of the most important recent biographies, *Darwin: The Life of a Tormented Evolutionist* by Adrian Desmond and James Moore, has been most helpful in establishing Darwin's attitudes about his work. The title is somewhat misleading: Darwin's tortures resulted mainly from ill health, overwork, and the death of his favorite daughter in 1850, not from his scientific conclusions. Also helpful has been *Darwin, His Daughter, and Human Evolution* by Randal Keynes.

11. Darwin, *Descent of Man*, 1:1.

12. Haraway, *Primate Visions*, 165. This collection of essays, along with Haraway's *Simians, Cyborgs, and Women*, has afforded me insights into and essential factual details about post-1950 primatology.

13. Quoted in Desmond and Moore, *Darwin*, 244.

14. Darwin, *Descent of Man*, 1:227.

15. Ibid., 1:11–12.

16. Ibid., 1:12.

17. Ibid., 1:192.

18. Ibid., 1:21.

19. Ibid., 2:277.

20. Ibid.

21. Appleman, *Darwin*, 180n. References to the second edition of *The Descent of Man* are from Philip Appleman's *Darwin: A Norton Critical Edition* (3rd ed.). This book is also useful in reconstructing the history of "Darwinism" after Darwin.

22. Darwin, *Descent of Man*, 2:371.

23. Ibid.

24. Ibid., 1:70.

25. Ibid., 1:71–72.

26. Ibid., 1:45.

27. Ibid., 1:78.

28. Ibid., 1:75–76.

29. Ibid., 1:140.

30. Ibid., 1:48.

31. Ibid., 2:389.

32. Ibid., 2:404–05.

33. Desmond and Moore, *Darwin*, 593.

34. Konrad Lorenz, introduction to Darwin, *Expression of the Emotions*, xii.

35. Darwin, *Expression of the Emotions*, 131.

36. Ibid., 144.

37. Ibid., 28.

38. Ibid., 50.

39. Ibid., 66.

40. Desmond and Moore, *Darwin*, 238.

41. Dawkins, *Selfish Gene*, 195–96. This discussion is also informed by Jared Diamond's *The Third Chimpanzee*.

Chapter 2

1. Dawkins, *Selfish Gene*.

2. Goodall, *Reason for Hope*, 11.

3. Cooper, *English Romance in Time*. Cooper's study is a useful recent analysis of quest romance, which, as an enormous body of literature, has received much literary critical attention.

4. Ibid., 66.

5. Mary Terrall mentions the "founding fathers" of astronomy as exemplars of this trope of the science adventurer—a category that was extended to include, in particular, the eighteenth-century French academicians Maupertuis and La Condamine, who wrote popular travel narratives about their geodetic explorations. Terrall, "Heroic Narratives of Quest and Discovery."

6. Lofting, *Voyages of Doctor Dolittle*, 71.

7. Ibid., 28.

8. Ibid., 47.

9. Ibid., 48–49.

10. Lofting, *Story of Doctor Dolittle*, 60.

11. Ibid., 79.

12. For more on the recent discoveries about the Dmanisi man, see the stories published on October 17, 2013, in multiple news outlets, including *National Geographic News*, the *Daily Mail*, the *Guardian*, and the *New York Times*.

13. The Leakey family's contributions to paleontology, anthropology, and primatology are so well known as to be summarized in almost every introductory text in all of these fields.

14. Goodall, *In the Shadow of Man*, 31.

15. The cover illustration of Haraway's *Primate Visions* and, especially, the article "Apes in Eden, Apes in Space" in that volume suggest the powerful iconographic impact of Goodall's first touch with David Graybeard.

16. Goodall, *In the Shadow of Man*, 38.

17. Ibid., 77, 47.

18. Goodall, *Through a Window*, 15.

19. Goodall, *In the Shadow of Man*, 270–71.

20. Goodall, *Through a Window*, 135.

21. De Waal, *Primates and Philosophers*, 112.

22. Goodall, *Through a Window*, 210.

23. The best summary I know of the Carpenter-Zuckerman episode in primatology appears in the first chapter of *Primate Encounters*, edited by Shirley Strum and Linda Marie Fedigan. They are the authors of this chapter, which serves as an introduction to the anthology.

24. Goodall, *Reason for Hope*, 117.

25. Ibid., 250.

26. The full title of Waldau and Patton's edited collection is *A Communion of Subjects: Animals in Religion, Science, and Ethics*.

27. Goodall, *Harvest for Hope*, xix.

Chapter 3

1. I have consulted Donna Haraway's *Primate Visions*; especially useful have been two essays from this volume: "A Pilot Plant for Human Engineering: Robert Yerkes and the Yale Laboratories of Primate Biology" and "Teddy Bear Patriarchy: Taxidermy in the Garden of Eden, New York City, 1908–36." See also Nash, "Gorilla Rhetoric."

2. Ugarte, *Shifting Ground*, 28.

3. See Vedder and Weber, *In the Kingdom of Gorillas*.

4. Schaller, "Gentle Gorillas, Turbulent Times."

5. Jenkins, "Who Murdered the Virunga Gorillas?"

6. Schaller, *Year of the Gorilla*, 21.

7. Ibid., 189.

8. Ibid., 179.

9. Mowat, *Woman in the Mists*, 1.

10. Fossey, *Gorillas in the Mist*, 5.

11. Fossey, "More Years with the Mountain Gorillas," 581.

12. Fossey, *Gorillas in the Mist*, 24–25.

13. Ibid., 166.

14. Prince-Hughes's *Songs of the Gorilla Nation* provides information on recent gorilla populations and distribution, as well as gorilla perception.

15. Fossey, *Gorillas in the Mist*, 141.

16. Fossey, "Making Friends with the Mountain Gorillas," 58, 60.

17. For further analysis on the touching of hands, see Karla Armbruster's article "'Surely, God, These Are My Kin': The Dynamics of Identity and Advocacy in the Life and Works of Dian Fossey."

18. Fossey, "Making Friends with the Mountain Gorillas," 63.

19. Fossey, "Imperiled Mountain Gorilla," 511.

20. Fossey, *Gorillas in the Mist*, 518.

21. Ibid., 522.

22. *Gorillas in the Mist* (1988), directed by Michael Apted, is a loose film adaptation of Fossey's book. Fossey was also a model for a remake of *Mighty Joe Young* (1998), directed by Ron Underwood, in which the lonely protagonist risks her life to save a giant gorilla, following in the footsteps of her mother, who was killed in the forest by poachers as she protected the gorillas. For Amy Vedder and Bill Weber in *In the Kingdom of Gorillas*, Fossey's death was the predictable end to a life of unbridled conflict and poor choices, but it is still a tragedy. Another recent telling of the story is Georgianne Nienaber's *Gorilla Dreams: The Legacy of Dian Fossey*, a novel narrated by the ghost of Fossey, who defends herself from the postmortem stories of monomania and bad science that have continued to circulate for more than twenty years. In this story, too, Fossey is tragically driven by conflicting impulses. As these stories and many more are told, death always seems the inevitable end to a life of privation, risk, conflict, mighty passions, and exile.

Chapter 4

1. In *Darwin's Plots*, Gillian Beer not only analyzes Darwin's literary strategies in great detail but also shows how he influenced the novelists of his day.

2. Wilson, "Foreword from the Scientific Side," ix.

3. Geertz, *Interpretation of Cultures*.

4. White, "Historical Text as Literary Artifact," 407.

5. Booth, "Ethics of Forms," 113, 126. Booth's essay was written for *Understanding Narrative*, edited by James Phelan and Peter J. Rabinowitz, whose introduction on narrative form is also helpful.

6. Cheney and Seyfarth, *How Monkeys See the World*, 48, 56. See also Cheney and Seyfarth, *Baboon Metaphysics*.

7. Strum, *Almost Human*, 39.

8. Ibid., 22.

9. Altmann, *Baboon Mothers and Infants*, 192–93.

10. Strum, *Almost Human*, 157.

11. Ibid., 161.

12. Ibid., 167.

13. The story continues in Hrdy's later works, *The Woman That Never Evolved*; *Mother Nature: Maternal Instincts and How They Shape the Human Species*; and *Mothers and Others*.

14. Bakhtin, "Discourse in the Novel," in *Dialogic Imagination*, 259–422. This is the longest essay in Bakhtin's book.

15. I am indebted to my colleague Phillip Lucas, who told me about seeing some of these behaviors on a recent trip to temples in India. Several articles about temple monkeys have recently been published in scientific journals, but one of the most accessible publications on this human-monkey association is Jane Teas's "Temple Monkeys of Nepal" in *National Geographic*, about rhesus macaques.

16. Hrdy, *Langurs of Abu*, 76–77.

17. Ibid., 77.

18. Ibid., 291.

19. Hrdy, *Mothers and Others*.

20. Hrdy, *Langurs of Abu*, 291.

21. Ibid., 1.

22. Ibid., 291.

23. Ibid., 309.

24. Hrdy, *Woman That Never Evolved*, 130.

25. Ibid., 129.

26. Rowell's review of *Almost Human* is quoted in "Praise for *Almost Human*" in the front fly-leaf of the 1987 paperback.

27. Bakhtin, *Dialogic Imagination*, 33.

Chapter 5

1. Strum, "Life with the Pumphouse Gang," photo on 682–83.

2. See Sue Savage-Rumbaugh's numerous publications on the language program with chimpanzees and bonobos, including her account of the famous bonobo, *Kanzi: The Ape at the Brink of the Human Mind*.

3. Strier, *Faces in the Forest*, 65, 85, 72. See also Strier, "American Primatologist Abroad in Brazil."

4. Strier, *Faces in the Forest*, xxx.

5. Ibid., xxvii.

6. Ibid., xxxviii–xxi.

7. Smuts, *Sex and Friendship in Baboons*, xvi.

8. Smuts, "Reflections," 109. Coetzee's *Lives of Animals* is such a foundational text in animal studies that scholars in this field now have to make an effort *not* to cite it! See also Smuts's "Encounters with Animal Minds" and "Embodied Communication in Non-human Animals."

9. Smuts, "Reflections," 110.

10. Ibid., 113. See also Haraway, "Apes in Eden, Apes in Space," in *Primate Visions*. Haraway reminds her readers that ever since Michelangelo painted the mutual touch of God and Adam on the Sistine Chapel ceiling, the representation of hands across a gulf of difference has resonated deeply. Smuts's description of the baboon's touch builds on this overdetermined motif.

11. Smuts, "Reflections," 113.

12. Ibid., 114.

13. Benjamin, "Desire of One's Own," 92.

14. Galdikas's approach leaves her open to highly charged negative criticism, of course. See, for example, *A Dark Place in the Jungle* by novelist Linda Spalding (Galdikas has sued Spalding for Spalding's representation of her and her conservation efforts). An academically oriented review charges Galdikas with failure to credit her colleagues and with giving disproportionate space to the facts of her own life (*Choice*, November 1995). On the other hand, *Kirkus* (December 1994) and *Booklist* (December 1994) reviewers praise the amount of information and detailed portraits of orangutans. *Publishers Weekly* (November 1994) treats *Reflections of Eden* as an adventure story.

15. Galdikas, *Reflections of Eden*, 125.

16. Ibid., 250.

17. Ibid., 255.

18. Ibid., 251.

19. Ibid., 5.

20. Ibid., 197.

21. Ibid.

22. This section on spiritual autobiography is informed by *The Female Autograph*, edited by Domna C. Stanton; see especially Stanton's own first chapter.

23. Galdikas, *Reflections of Eden*, 403.

24. For accessible recent accounts see Deborah Blum's *The Monkey Wars* and *Love at Goon Park*.

Chapter 6

1. Sapolsky, *Primate's Memoir*, 140.

2. "How I Write."

3. Ibid.

4. Sapolsky, *Primate's Memoir*, 13.

5. Bakhtin, *Rabelais and His World*.

6. Sapolsky, *Monkeyluv*, 176.

7. Sapolsky, *Primate's Memoir*, 20.

8. Ibid., 99.

9. Ibid., 24.

10. Ibid.

11. Ibid., 39.

12. Laurence Frank has published articles about hyena behavior in *Science*, *Animal Behavior*, and other venues.

13. Sapolsky, *Primate's Memoir*, 157.

14. Ibid., 220.

15. Ibid., 222.

16. Ibid.

17. Ibid., 227.

18. Ibid., 230.

19. Ibid., 276.

20. Ibid., 301.

21. Ibid., 302–3.

22. Ibid., 303.

23. For more recent history of Sapolsky's study troop, see Natalie Angier's article on Sapolsky's work, "No Time for Bullies: Baboons Retool Their Culture," and Sapolsky's own "Warrior Baboons Give Peace a Chance."

Conclusion

1. Kummer, *In Quest of the Sacred Baboon*, 82–83.

2. For more information on the Lemur Conservation Foundation and the Myakka City Lemur Reserve, see the webpage "Myakka City Lemur Reserve" (http://www.lemurreserve.org/myakka.html). A happy irony: the Lemur Conservation Foundation and the reserve were founded and initially funded by Penelope Bodry-Sanders, the author of two biographies of the early twentieth-century hunter and taxidermist Carl Akeley, whose mountain gorillas dominate the African Hall in the American Museum of Natural History. Akeley's remorse about killing the animals inspired him to begin the political process of establishing the Virungas as a park for their protection. Bodry-Sanders's lemur project brings Akeley's colonial extractionist adventuring full circle, with hope to reintroduce species to their native habitat.

3. Burroughs, *Ghost of Chance*, 15.

4. Quammen, *Song of the Dodo*, 506. This book is a brilliant journalistic investigation of island biogeography and extinction.

5. For the observation that lemurs have never been studied as members of intact ecological communities, see Alison F. Richard, "Malagasy Prosimians." See also Jolly, *World Like Our Own*, and the IUCN publication *Key Environments: Madagascar*, edited by Alison Jolly, Philippe Oberlé, and Roland Albignac. Jolly's *National Geographic* article "Madagascar's Lemurs on the Edge of Survival" also serves as a good primer on lemurs.

6. The discovery of Ida, the extinct primate fossil, was announced in May 2009 by a team of paleontologists and received widespread media coverage.

7. Schaller, "Gentle Gorillas, Turbulent Times." This is one of a cluster of short, related articles featuring Michael Nichols's photography.

8. Jenkins, "Who Murdered the Virunga Gorillas?"

9. Strier presented these findings in the lecture "Endangered Muriqui Monkeys of Brazil."

10. Byatt, *Possession*, 259.

11. Laura Marsh's *Primates in Fragments* is a pioneering anthology in fragmentation science, written from the perspective of primatology.

12. Jolly, *Lords and Lemurs*, 274–75.

13. Woods, *Bonobo Handshake*, 32.

14. Marsh, *Primates in Fragments*, 371.

BIBLIOGRAPHY

Altmann, Jeanne. *Baboon Mothers and Infants*. Cambridge: Harvard University Press, 1980.

Angier, Natalie. "No Time for Bullies: Baboons Retool Their Culture." *New York Times*, April 13, 2004, F1–2.

Appleman, Philip, ed. *Darwin: A Norton Critical Edition*. 3rd ed. New York: W. W. Norton, 2001.

Armbruster, Karla. "'Surely, God, These Are My Kin': The Dynamics of Identity and Advocacy in the Life and Works of Dian Fossey." In *Animal Acts: Configuring the Human in Western History*, edited by Jennifer Hamm and Matthew Senior, 209–29. New York: Routledge, 1997.

Bakhtin, Mikhail, ed. *The Dialogic Imagination: Four Essays*. Translated by Michael Holquist and Caryl Emerson. Austin: University of Texas Press, 1981.

———. *Rabelais and His World*. Translated by Hélène Iswolsky. Bloomington: Indiana University Press, 1984.

Beer, Gillian. *Darwin's Plots: Evolutionary Narrative in Darwin, George Eliot, and Nineteenth-Century Fiction*. London: Routledge and Kegan Paul, 1983.

Benjamin, Jessica. "A Desire of One's Own: Psychoanalytic Feminism and Intersubjective Space." In *Feminist Studies / Critical Studies*, edited by Teresa de Lauretis, 78–101. Bloomington: Indiana University Press, 1986.

Bloom, Harold. *The Visionary Company: A Reading of English Romantic Poetry*. Ithaca: Cornell University Press, 1971.

Blum, Deborah. *Love at Goon Park: Harry Harlow and the Science of Affection*. Cambridge: Perseus, 2002.

———. *The Monkey Wars*. New York: Oxford University Press, 1994.

Booth, Wayne. "The Ethics of Forms: Taking Flight with *The Wings of a Dove*." In *Understanding Narrative*, edited by James Phelan and Peter J. Rabinowitz, 99–135. Columbus: Ohio State University Press, 1994.

Burroughs, William S. *Ghost of Chance*. London: Serpent's Tale Press, 1995.

Byatt, A. S. *Possession*. New York: Random House, 1991.

Cheney, Dorothy L., and Robert M. Seyfarth. *Baboon Metaphysics: The Evolution of a Social Mind*. Chicago: University of Chicago Press, 2007.

———. *How Monkeys See the World: Inside the Mind of Another Species*. Chicago: University of Chicago Press, 1992.

Cooper, Helen. *The English Romance in Time: Transforming Motifs from Geoffrey of Monmouth to the Death of Shakespeare*. New York: Oxford University Press, 2004.

Darwin, Charles. *The Descent of Man, and Selection in Relation to Sex*. Facsimile ed. Princeton: Princeton University Press, 1981.

———. *The Expression of the Emotions in Man and Animals*. Facsimile ed. Introduction by Konrad Lorenz. Chicago: University of Chicago Press, 1965.

———. *The Origin of Species*. 6th ed. Introduction by Julian Huxley. New York: New American Library, 1958.

Dawkins, Richard. *The Selfish Gene*. New York: Oxford University Press, 1976.

Desmond, Adrian, and James Moore. *Darwin: The Life of a Tormented Evolutionist*. New York: W. W. Norton, 1991.

de Waal, Frans. *Our Inner Ape: A Leading Primatologist Explains Why We Are Who We Are*. New York: Penguin, 2005.

———. *Primates and Philosophers: How Morality Evolved*. Edited by Stephen Macedo and Josiah Ober. Princeton: Princeton University Press, 2006.

Diamond, Jared. *The Third Chimpanzee: The Evolution and Future of the Human Animal*. New York: Harper Perennial, 1992.

Fossey, Dian. *Gorillas in the Mist*. Boston: Houghton Mifflin, 1983.

———. "The Imperiled Mountain Gorilla." *National Geographic Magazine*, April 1981, 501–23.

———. "Making Friends with the Mountain Gorillas." *National Geographic Magazine*, January 1970, 48–67.

———. "More Years with the Mountain Gorillas." *National Geographic Magazine*, October 1971, 574–85.

Galdikas, Birute M. F. "Living with the Great Orange Apes." *National Geographic Magazine*, June 1980, 830–53.

———. *Orangutan Odyssey*. New York: Harry N. Abrams, 1999.

———. "Orangutans, Indonesia's 'People of the Forest.'" *National Geographic Magazine*, October 1975, 444–73.

———. *Reflections of Eden: My Years with the Orangutans of Borneo*. New York: Little, Brown, 1995.

Geertz, Clifford. *The Interpretation of Cultures: Selected Essays by Clifford Geertz*. New York: Basic Books, 1973.

Gilbert, Sandra, and Susan Gubar. *The Madwoman in the Attic: The Woman Writer and the Nineteenth-Century Imagination*. New Haven: Yale University Press, 1979.

Goodall, Jane. *The Chimpanzees of Gombe*. Cambridge: Harvard University Press, 1986.

———. *Harvest for Hope: A Guide to Mindful Eating*. With Gary McAvoy and Gail Hudson. New York: Warner Books, 2005.

———. *Hope for Animals and Their World: How Endangered Species Are Being Rescued from the Brink*. New York: Hachette Book Group, 2009.

———. *In the Shadow of Man.* New York: Dell, 1971.

———. *Reason for Hope: A Spiritual Journey.* With Phillip Berman. New York: Warner Books, 1999.

———. *Through a Window: My Thirty Years with the Chimpanzees of Gombe.* Boston: Houghton Mifflin, 1990.

Griffin, Donald R. *Animal Thinking.* Cambridge: Harvard University Press, 1984.

Haraway, Donna. *The Haraway Reader.* New York: Routledge, 2004.

———. *Primate Visions: Gender, Race, and Nature in the World of Modern Science.* New York: Routledge, 1989.

———. *Simians, Cyborgs, and Women: The Reinvention of Nature.* New York: Routledge, 1991.

Harcourt, Caroline S., and Jane Thornback. *Lemurs of Madagascar and the Comoros: The IUCN Red Data Book.* Cambridge: World Conservation Union, 1990.

Heise, Ursala. *Sense of Place and Sense of Planet: The Environmental Imagination of the Global.* New York: Oxford University Press, 2008.

Hrdy, Sarah Blaffer. *The Langurs of Abu: Female and Male Strategies of Reproduction.* Cambridge: Harvard University Press, 1980.

———. *Mother Nature: Maternal Instincts and How They Shape the Human Species.* New York: Ballantine Books, 1999.

———. *Mothers and Others: The Evolutionary Origins of Mutual Understanding.* Cambridge: Harvard University Press, 2009.

———. *The Woman That Never Evolved.* Cambridge: Harvard University Press, 1981.

Huxley, Thomas. *Man's Place in Nature.* New York: Appleton, 1896.

Jenkins, Mark. "Who Murdered the Virunga Gorillas?" *National Geographic Magazine*, July 2008, 34–65.

Jolly, Alison. Email to teresopolis@majordomo.srv.ualberta.edu. In *Primate Encounters: Models of Science, Gender, and Society*, edited by Shirley Strum and Linda Marie Fedigan, 537. Chicago: University of Chicago Press, 2000.

———. *Lords and Lemurs: Mad Scientists, Kings with Spears, and the Survival of Diversity in Madagascar.* Boston: Houghton Mifflin, 2004.

———. "Madagascar's Lemurs on the Edge of Survival." *National Geographic Magazine*, August 1988, 132–61.

———. *A World Like Our Own: Man and Nature in Madagascar.* New Haven: Yale University Press, 1980.

Jolly, Alison, Philippe Oberlé, and Roland Albignac, eds. *Key Environments: Madagascar.* New York: Pergamon Press, 1984.

Kennedy, J. Gerald, and Liliane Weissberg, eds. *Romancing the Shadow: Poe and Race.* New York: Oxford University Press, 2001.

Keynes, Randal. *Darwin, His Daughter, and Human Evolution.* New York: Penguin, 2002.

Korsgaard, Christine M. "Morality and the Distinctiveness of Human Action." In *Primates and Philosophers: How Morality Evolved*, edited by Stephen Macedo and Josiah Ober, 98–119. Princeton: Princeton University Press, 2006.

Kuhn, Thomas S. *The Structure of Scientific Revolutions*. Chicago: University of Chicago Press, 1970.

Kummer, Hans. *In Quest of the Sacred Baboon: A Scientist's Journey*. Princeton: Princeton University Press, 1995.

Levine, George. *Darwin Loves You: Natural Selection and the Re-enchantment of the World*. Princeton: Princeton University Press, 2006.

Lofting, Hugh. *The Story of Doctor Dolittle*. New York: Doubleday, 1988.

——. *The Voyages of Doctor Dolittle*. New York: F. A. Stokes, 1922.

Marais, Eugène N. *The Soul of the Ape*. Introduction by Robert Ardrey. Harmondsworth, Middlesex: Penguin, 1973.

Marsh, Laura K., ed. *Primates in Fragments: Ecology in Conservation*. New York: Kluwer Academic / Plenum, 2003.

Martin, Wallace. *Recent Theories of Narrative*. Ithaca: Cornell University Press, 1986.

McEwan, Ian. "Evolution and Literary Theory." In *The Literary Animal: Evolution and the Nature of Narrative*, edited by Jonathan Gottschall and David Sloan Wilson, 5–19. Evanston: Northwestern University Press, 2005.

Mowat, Farley. *Woman in the Mists: The Story of Dian Fossey and the Mountain Gorillas of Africa*. New York: Warner Books, 1987.

Nash, Richard. "Gorilla Rhetoric: Family Values in the Mountains." *Symplokē* 4, nos. 1–2 (1996): 95–133.

Nienaber, Georgianne. *Gorilla Dreams: The Legacy of Dian Fossey*. Lincoln, Neb.: iUniverse, 2006.

Poe, Edgar Allan. "The Murders in the Rue Morgue." In *Poetry and Tales*, edited by Patrick F. Quinn, 397–431. Library of America Series. New York: Literary Classics of America, 1984.

Prince-Hughes, Dawn. *Songs of the Gorilla Nation: My Journey Through Autism*. New York: Random House, 2004.

Quammen, David. *Song of the Dodo: Island Biogeography in an Age of Extinction*. New York: Simon and Schuster, 1996.

Rich, Adrienne. "The Observer." In *The Fact of a Door Frame: Poems Selected and New, 1950–1984*, 97–98. New York: W. W. Norton, 1984.

Richard, Alison F. "Malagasy Prosimians: Female Dominance." In *Primate Societies*, edited by Barbara Smuts, Dorothy Cheney, Robert Seyfarth, Richard Wrangham, and Thomas Struhsaker. Chicago: University of Chicago Press, 1987.

Sapolsky, Robert M. "How I Write—Conversation Transcript." web.stanford .edu/group/howiwrite/Transcripts/Sapolsky_transcript.html.

———. *Monkeyluv and Other Essays on Our Lives as Animals.* New York: Scribner, 2005.

———. *A Primate's Memoir: A Neuroscientist's Unconventional Life Among the Baboons.* New York: Touchstone, 2001.

———. "Warrior Baboons Give Peace a Chance." *YES!*, Spring 2011, 34–37.

———. *Why Zebras Don't Get Ulcers.* 3rd ed. New York: Henry Holt, 2004.

Savage-Rumbaugh, Sue. *Kanzi: The Ape at the Brink of the Human Mind.* New York: John Wiley and Sons, 1994.

Schaller, George B. "Gentle Gorillas, Turbulent Times." *National Geographic Magazine*, October 1995, 65–71.

———. *The Year of the Gorilla.* Chicago: University of Chicago Press, 1964.

Shipman, Pat. *The Animal Connection: A New Perspective on What Makes Us Human.* New York: W. W. Norton, 2011.

Smith, Barbara Hernnstein. *Scandalous Knowledge: Science, Truth, and the Human.* Durham: Duke University Press, 2005.

Smuts, Barbara. "Between Species: Science and Subjectivity." *Configurations* 14, no. 1 (2006): 115–26.

———. "Embodied Communication in Non-human Animals." In *Human Development in the Twenty-First Century: Visionary Ideas from Systems Scientists*, edited by Alan Fogel, Barbara J. King, and Stuart G. Shanker, 136–46. New York: Cambridge University Press, 2007.

———. "Encounters with Animal Minds." *Journal of Consciousness Studies* 8, nos. 5–7 (2001): 293–309.

———. "Reflections." In *The Lives of Animals*, by J. M. Coetzee, 107–20. Princeton: Princeton University Press, 1999.

———. *Sex and Friendship in Baboons.* 2nd ed. Cambridge: Harvard University Press, 1999.

Spalding, Linda. *A Dark Place in the Jungle: Following Leakey's Last Angel into Borneo.* Chapel Hill: Algonquin Books, 2003.

Stanton, Domna C., ed. *The Female Autograph: Theory and Practice of Autobiography from the Tenth to the Twentieth Century.* Chicago: University of Chicago Press, 1984.

Strier, Karen B. "An American Primatologist Abroad in Brazil." In *Primate Encounters: Models of Science, Gender, and Society*, edited by Shirley C. Strum and Linda Marie Fedigan, 194–207. Chicago: University of Chicago Press, 2000.

———. "The Endangered Muriqui Monkeys of Brazil: Science, Gender, and Conservation." Lecture, Stetson University, DeLand, Fla., October 30, 2008.

———. *Faces in the Forest: The Endangered Muriqui Monkeys of Brazil.* 2nd ed. Cambridge: Harvard University Press, 1999.

Strum, Shirley. *Almost Human: A Journey into the World of Baboons.* New York: W. W. Norton, 1987.

———. "Life with the Pumphouse Gang." *National Geographic Magazine,* May 1975, 672–91.

Strum, Shirley, and Linda Marie Fedigan, eds. *Primate Encounters: Models of Science, Gender, and Society.* Chicago: University of Chicago Press, 2000.

Teas, Jane. "Temple Monkeys of Nepal." *National Geographic Magazine,* April 1980, 574–84.

Terrall, Mary. "Heroic Narratives of Quest and Discovery." In *The Postcolonial Science and Technology Studies Reader,* edited by Sandra Harding, 84–102. Durham: Duke University Press, 2011.

Ugarte, Michael. *Shifting Ground: Spanish Civil War Exile Literature.* Durham: Duke University Press, 1989.

Vedder, Amy, and Bill Weber. *In the Kingdom of Gorillas: Fragile Species in a Dangerous Land.* New York: Simon and Schuster, 2001.

Waldau, Paul, and Kimberley Patton, eds. *A Communion of Subjects: Animals in Religion, Science, and Ethics.* New York: Columbia University Press, 2006.

Weil, Kari. *Thinking Animals: Why Animal Studies Now?* New York: Columbia University Press, 2012.

White, Hayden. "The Historical Text as Literary Artifact." In *Critical Theory Since 1965,* edited by Hazard Adams and Leroy Searle, 395–407. Florida: University Press of Florida, 1986.

Wilson, E. O. "Foreword from the Scientific Side." In *The Literary Animal: Evolution and the Nature of Narrative,* edited by Jonathan Gottschall and David Sloan Wilson, vii–xi. Evanston: Northwestern University Press, 2005.

Woods, Vanessa. *Bonobo Handshake: A Memoir of Love and Adventure in the Congo.* New York: Gotham Books, 2010.

———. *It's Every Monkey for Themselves: A True Story of Sex, Love, and Lies in the Jungle.* Sydney: Allen and Unwin, 2007.

Yaeger, Patricia. *Honey-Mad Women: Emancipatory Strategies in Women's Writing.* New York: Columbia University Press, 1988.

Yerkes, Robert, and Ada Yerkes. *The Great Apes.* New Haven: Yale University Press, 1929.

INDEX